Radiotherapy

and

Clinical Radiobiology

of

Head and Neck Cancer

Series in Medical Physics and Biomedical Engineering

Series Editors: John G. Webster, E. Russell Ritenour, Slavik Tabakov,
and Kwan-Hoong Ng

Recent books in the series:

Radiotherapy
and
Clinical Radiobiology
of
Head and Neck Cancer

Loredana G. Marcu
Iuliana Toma-Dasu
Alexandru Dasu
Claes Mercke

CRC Press
Taylor & Francis Group
Boca Raton London New York

CRC Press is an imprint of the
Taylor & Francis Group, an **informa** business

CRC Press
Taylor & Francis Group
6000 Broken Sound Parkway NW, Suite 300
Boca Raton, FL 33487-2742

Library of Congress Cataloging-in-Publication Data

Names: Marcu, Loredana, author. | Toma-Dasu, Iuliana, author. | Dasu, Alexandru, author. | Mercke, Claes, 1942- author.
Title: Radiotherapy and clinical radiobiology of head and neck cancer / by Loredana G. Marcu, IulianaToma-Dasu, Alexandru Dasu, Claes Mercke.
Description: Boca Raton, FL : CRC Press, Taylor & Francis Group, [2018] |
Series: Series in medical physics and biomedical engineering | Includes bibliographical references and index.
Identifiers: LCCN 2018003608 (print) | LCCN 2018017753 (ebook) | ISBN 9781351002004 (eBook General) | ISBN 9781351001991 (eBook PDF) | ISBN 9781351001984 (eBook ePub) | ISBN 9781351001977 (eBook Mobipocket) | ISBN 9781498778299 | ISBN 9781498778299 (hardback ;alk. paper) | ISBN 1498778291 (hardback ;alk. paper)
Subjects: LCSH: Head--Cancer--Radiotherapy. | Neck--Cancer--Radiotherapy.
Classification: LCC RC280.H4 (ebook) | LCC RC280.H4 M28 2018 (print) | DDC 616.99/4910642--dc23
LC record available at https://lccn.loc.gov/2018003608

Visit the Taylor & Francis Web site at
http://www.taylorandfrancis.com

and the CRC Press Web site at
http://www.crcpress.com

Contents

Foreword

This new book on the Radiotherapy and Clinical Radiobiology of Head and Neck Cancer covers a wide range of relevant topics. These include risk factors and pathogenesis; cancer anatomy trends; history of treatments; radiobiology refresher; hypoxia and angiogenesis; tumour cell repopulation; HPV-associated cancers; normal tissue tolerance; radiotherapy modalities; targeted therapies; retreatment issues; and clinical radiobiology modelling of schedules. It was written by Loredana Marcu, Iuliana Toma-Dasu, and Alexandru Dasu, who have expertise in medical physics and cancer research; and senior radiation oncologist, Claes Mercke. Many references are given for further information on particular aspects of the subject.

Over half a million new cases of head and neck cancer (HNC) are reported annually worldwide, representing 2%–7% of all cancers in various national or regional registries. Radiotherapy has been used successfully for many decades for treating HNC, because of the accessibility of those tumours in the body, the clear outcome criteria and the dominant importance of locoregional control for survival. This resulted in much clinical radiobiology modelling of radiotherapy dose-response and schedule dependency on outcomes being established for HNC before some more common types of cancer. Survival rates have increased over many years. This has been due to technology innovations, such as intensity-modulated and image-guided radiotherapy, schedule modifications like hyper- and accelerated radiotherapy, and combination treatments with chemotherapy and targeted agents.

There are several books available that cover some of the aspects described here. In the last 10 years, the multi-author/chapter volumes have included, *Combining Targeted Biological Agents with Radiotherapy: Current Status and Future Directions*, (editor William Small, 2008), *Head and Neck Cancer*, (editors Bruce Brockstein, Gregory Masters, 2010), *Head and Neck Cancer; Multimodality Management*, (editor Jacques Bernier, 2016), and *Fundamentals of Radiotherapy in Head and Neck Cancer*, (editors Samta Mittal and Suraj Agarwal, 2016). However, this volume contains more information on the present state of knowledge on the clinical radiobiology of the disease and its treatment. Also, it is always helpful to have up-to-date text on the subject. This book should appeal, in particular, to trainee radiation oncologists who are joining a head and neck team and requires a concise description of the science underpinning existing and developing radiotherapy practices for this type of cancer. In summary, I recommend this new book to all trainees and practicing radiation oncologists, medical physicists, and clinical radiobiologists, who are involved in various ways with HNC patients and their treatment outcomes.

Professor Jolyon H Hendry
Manchester, UK

Preface

Writing a scientific book is always challenging due to the desire to entice the reader by offering something different and stimulating, in a manner that will help the reader to grasp new concepts while also revisiting traditional views. The idea of this book came about because there was a need to conceptually join radiobiology and radiotherapy of head and neck cancer (HNC) to illustrate the advances in treatment along with developments in molecular biology.

Over the past few decades, there were several landmark discoveries that have shaped the treatment of HNC. These landmarks, which are frequently related to radiobiology, have dictated the development of new treatments and improvements in the existing ones. As the science behind radiotherapy, radiobiology justifies the clinical implementation of various treatment techniques and schedules, and at the same time, gives indications regarding tumour control and normal tissue effects after treatment. This leads to HNCs being one of the tumours with the most radiobiological insights for the quantifiable response to radiation therapy.

HNCs are radiobiologically challenging due to several factors that strongly impact treatment outcome: tumour hypoxia, the subpopulation of cancer stem cells, accelerated tumour repopulation during treatment, HPV status, etc. Advances in understanding of the aforementioned factors and their role in guiding tumour behaviour are crucial aspects of today's cancer management. While locoregional control of HNC has improved over the years, overall survival among HNC patients is yet to become a statistic of which clinicians are proud. The systemic management of HNC and the retreatment of patients are other challenging aspects that require an interdisciplinary solution.

All of the chapters that describe the radiobiological aspects of HNC also discuss their clinical implications. The authors aimed to link the scientific rationale to the clinical management of HNC and to justify the success (or failure) of various trials or clinical studies.

In view of the prominent place of radiation therapy in the management of HNC, we consider that this book contributes to the education of radiation therapy professionals, especially in consolidating the knowledge on factors influencing the radiation response of head and neck tumours and normal tissues, and as the basis for exploring new treatment techniques. Through its interdisciplinary approach, this book should be of interest to radiation oncologists, clinical oncologists, radiation biologists, radiation therapists, and medical physicists alike.

The journey of book writing cannot be undertaken without some guidance and support from the surrounding environment. With this in mind, the authors would like to thank their families for their endless patience, encouragement, and support; and for believing in the "light at the end of the tunnel."

A warm thank you is reserved for two special women from CRC Press: Francesca McGowan, who persuaded us to start this project; and Rebecca Davies, who followed our progress with great dedication and guided us through the publishing process.

Loredana G. Marcu
Iuliana Toma-Dasu
Alexandru Dasu
Claes Mercke

Series Preface

ABOUT THE SERIES

The Series in Medical Physics and Biomedical Engineering describes the applications of physical sciences, engineering, and mathematics in medicine and clinical research.

The series seeks (but is not restricted to) publications in the following topics:

- Artificial organs
- Assistive technology
- Bioinformatics
- Bioinstrumentation
- Biomaterials
- Biomechanics
- Biomedical engineering
- Clinical engineering
- Imaging
- Implants
- Medical computing and mathematics
- Medical/surgical devices

- Patient monitoring
- Physiological measurement
- Prosthetics
- Radiation protection, health physics, and dosimetry
- Regulatory issues
- Rehabilitation engineering
- Sports medicine
- Systems physiology
- Telemedicine
- Tissue engineering
- Treatment

The *Series in Medical Physics and Biomedical Engineering* is an international series that meets the need for up-to-date texts in this rapidly developing field. Books in the series range in level from introductory graduate textbooks and practical handbooks to more advanced expositions of current research.

The *Series in Medical Physics and Biomedical Engineering* is the official book series of the International Organization for Medical Physics.

THE INTERNATIONAL ORGANIZATION FOR MEDICAL PHYSICS

The International Organization for Medical Physics (IOMP) represents over 18,000 medical physicists worldwide and has a membership of 80 national and 6 regional organizations, together with a number of corporate members. Individual medical physicists of all national member organizations are also automatically members.

The IOMP's mission is to advance medical physics practice worldwide by disseminating scientific and technical information, fostering the educational and professional development of medical physics, and promoting the highest quality medical physics services for patients.

A World Congress on Medical Physics and Biomedical Engineering is held every three years in cooperation with the International Federation for Medical and Biological Engineering (IFMBE) and the International Union for Physics and Engineering Sciences in Medicine (IUPESM). A regionally based international conference, the International Congress of Medical Physics (ICMP), is held between world congresses. In addition, the IOMP sponsors international conferences, workshops, and courses.

The IOMP has several programs to assist medical physicists in developing countries. The joint IOMP Library Program supports 75 active libraries in 43 developing countries, and the Used Equipment Program coordinates equipment donations. The Travel Assistance Program provides a limited number of grants to enable physicists to attend the world congresses. The IOMP co-sponsors the *Journal of Applied Clinical Medical Physics* and twice a year, it publishes an electronic bulletin, Medical Physics World. The IOMP also publishes *e-Zine*, an electronic newsletter, about six times a year. Additionally, the IOMP has an agreement with Taylor & Francis for the publication of the *Medical Physics and Biomedical Engineering* series of textbooks and members of the IOmP receive a discount.

IOMP collaborates with international organizations, such as the World Health Organization (WHO), the International Atomic Energy Agency (IAEA), and other international professional bodies—such as the International Radiation Protection Association (IRPA) and the International Commission on Radiological Protection (ICRP)—to promote the development of medical physics and the safe use of radiation and medical devices.

Guidance on education, training, and professional development of medical physicists is issued by IOMP, which collaborates with other professional organizations to develop a professional certification system for medical physicists that can be implemented globally.

The IOMP website (www.iomp.org) contains information on all the activities of the IOMP, Policy Statements 1 and 2, and the *IOMP: Review and Way Forward*, which outlines all its current activities and its plans for the future.

Authors

Loredana G. Marcu is Professor of Medical Physics at the University of Oradea, Romania, and Adjunct Professor at School of Health Sciences, University of South Australia. She received her PhD in Medical Physics from the University of Adelaide, and her Masters degree in Applied Physics from the West University of Timisoara, Romania. During her experience in Australia, she coordinated the low dose rate brachytherapy programme at the Royal Adelaide Hospital. She was also a Training Education and Accreditation Program (TEAP) preceptor supervising and coordinating the medical physics training and education of junior physicists in South Australia.

Her 20 years of teaching experience at both Australian and Romanian universities has culminated in 14 books and book chapters on physics, radiobiology, and teaching methodologies. Her current research interests cover in silico modelling of tumour growth and response to treatment, targeted therapies, the radiobiology of head and neck cancer (HNC), and the risk of secondary cancer after radiotherapy. Dr Marcu is the recipient of the 'Boyce Worthley award 2006', which is awarded by the Australasian College of Physical Scientists and Engineers in Medicine for her achievements in the areas of radiobiology and medical physics.

Iuliana Toma-Dasu is Associate Professor in Medical Radiation Physics in the Department of Physics, Stockholm University, and she is the research group leader of the Stockholm University Medical Radiation Physics division, which is affiliated with the Department of Oncology and Pathology at the Karolinska Institutet.

Dr Toma-Dasu studied Medical Physics at Umeå University, Sweden, and received a PhD degree in 2004. Her PhD studies were led by two prominent names in the field of radiobiology, Prof Juliana Denekamp and Prof Jack Fowler. She became a certified medical physicist in 2005 and joined the Medical Radiation Physics group at the Karolinska Institutet in Stockholm. Since 2010, she has been Associate Professor in Medical Radiation Physics at Stockholm University and the leader of the Medical Radiation Physics Division at Stockholm University and the Karolinska Institutet since 2013. In parallel with her heavy involvement in the educational programme for medical physicists at Stockholm University, her main research interests focus on exploring the influence of the tumour microenvironment on treatment outcome and the biological optimisation and adaptation of radiation therapy based on functional imaging.

Alexandru Dasu is the Chief Medical Physicist at the Skandion Clinic, the national Swedish proton centre in Uppsala and Associate Professor in Medical Radiation Physics.

Dr Dasu studied Medical Physics at Umeå University, Sweden, and received a PhD degree in 2001. He became a certified medical physicist in 2003 and Associate Professor in Medical Radiation Physics in 2008. He has synergistically combined clinical practice in medical radiation physics with top research work in the field of radiotherapy. In addition to his strong clinical and research interest in proton therapy, another one of his major research interests is the modelling of the influence of tumour microenvironment on the tumour response to radiation with special emphasis on the tumour oxygenation, the study of the response of tumours to various fractionated regimens in relation to their radiobiological parameters, and the risk for stochastic effects following radiotherapy, including the risk of secondary cancers.

Claes Mercke is Professor of Radiation Oncology and one of the senior physicians at the Karolinska University Hospital, Department of Oncology, in Stockholm, Sweden. He is a member of a team of physicians who are responsible for the non-surgical treatment programmes of patients with malignant diseases in the head and neck, thorax, and skin and is also responsible for the radiotherapy of paediatric tumours.

Dr Mercke studied medicine in Lund and Stockholm, Sweden, and was a certified specialist first in medicine and subsequently in general oncology. He received his PhD in Oncology and Medicine from the University of Lund and was soon appointed associate professor, where he was affiliated with the same university. He was head of the division of radiotherapy at the Department of Oncology at Sahlgren's University Hospital in Gothenburg for about 20 years but moved to Stockholm and the Karolinska University Hospital in 2006 after being appointed Professor of Radiation Oncology at Karolinska Institutet. His main research focus has been on radiobiology, especially with a focus on genetically determined radiosensitivity as applied to clinical radiotherapy. Dr Mercke has implemented various techniques, such as modern high dose rate and pulsed dose rate brachytherapy and IMRT for HNC, prostate cancer, and paediatric tumours. Teaching and the supervision of students for their PhD exams has been one of Dr Mercke's important responsibilities for a long time.

CHAPTER 1

Introductory Aspects of Head and Neck Cancers

1.1 HISTORIC ASPECTS

Squamous cell carcinoma of the head and neck region is a heterogeneous group of malignant diseases. Tumours arise from many anatomic subsites in the region and tumours with this same histology originate from the mucous membranes in the area. The head and neck region is a challenging region of the body, from a therapeutic point of view, since so many parts of it are associated with important basic physiologic functions such as nutrition, respiration, and communication, including speech. Severe debilitating consequences may occur as a result of tumours in this region, depending on specific location, size, and spread of pattern of the tumour. Structural disfigurement and functional impairments may compromise social integration and quality of life considerably.

For long, *surgical treatment* with an attempt at radical resection of the tumour was the dominant approach to treat these tumours. In 1905, by George W. Crile stated: 'Owing to the fact that no other treatment has offered as much as the operative…that there has been a hitherto high mortality and low permanent cure rate and that in many instances death comes in its most horrible form, the surgical treatment of cancer of the head and neck has ever taxed the courage and the ingenuity of the surgeon' (Crile 1906).

However, traditional approaches with surgical resections have, despite advances in the past, often resulted in significant morbidities, at least for patients with advanced tumours. The discovery of radium in Paris by Marie and Pierre Curie in 1898 started the evolution of *nonsurgical therapy* for head and neck cancer in the form of *radiotherapy* and the first successful treatments of two basal cell cancers in the face were reported in 1903, where the treating teams used brachytherapy. This technique of radiotherapy began almost immediately after the discovery of X-rays by Röntgen in 1895 and the discovery by Becquerel in 1896 of the natural radiation emissions from uranium crystals. Subsequently, there was a host of developments in this sphere of radiotherapy but with primitive equipment and dosimetry and with manually preloaded sources. In the early 1930s, more refined dosimetry systems made therapy much more precise when the Manchester system was devised together with the Paterson-Parker rules of implantation. These important inventions for the execution of radiotherapy in the form of brachytherapy were soon entered into general, clinical practice; using moulds, intracavitary, intraluminal, or interstitial techniques in a remote afterloading fashion.

However, from the 1950s onward, rapid technical evolution of external radiotherapy made the practice of brachytherapy for head and neck cancer relatively uncommon. The high-energy teletherapy machines with sophisticated treatment planning systems, dosimetric methods, improved good-depth dose delivery, and radiation safety made indications for interstitial radium brachytherapy to decrease significantly. With the discovery of artificial radioactivity by Irene Curie and Frederic Joliot in 1933, brachytherapy later staged a revival for certain patients with head and neck cancers. During subsequent decades, radiotherapy, most used in the form of external beam radiotherapy, emerged as a cornerstone for the treatment of patients with squamous cell carcinoma of the head and neck, either used alone or in a multimodality setting with drugs or with surgery (Thames 1988; Corvo 2007; Lacas 2017).

After World War II, the use of folic acid derivatives was demonstrated for the treatment of leukaemia. This finding, that antimetabolites could inhibit tumour growth through the metabolic inhibition of DNA synthesis, later served as the basis for the first *chemotherapy* treatments of head and neck cancer patients with methotrexate. The introduction of cisplatin and later other platinum analogues first for testicular carcinoma and subsequently for several other malignancies was of great importance and also for the treatment of patients with squamous cell carcinomas of the head and neck. Important results have been achieved with several combination chemotherapy regimens, preferably for patients with recurrent tumours but also for patients with primary, untreated locoregionally advanced tumours, in combination with radiotherapy. Platinum-containing regimens form the basis of treatment for most patients with recurrent, nonresectable tumours, after radiotherapy. Moreover, cisplatin administered concomitantly with curative radiotherapy for locoregionally advanced tumours has increased 5-year survival with at least 5%. Along with altered fractionation in the form of hyperfractionation, concomitant chemoradiotherapy is now a standard of care for the treatment of locally advanced squamous cell cancers (Budach 2006; Blanchard 2011; Lacas 2017). The last decade has also observed interest in targeted therapies for head and neck cancer; in 2006, cetuximab was approved by the FDA for the treatment of patients with recurrent/metastatic tumours as well as an addition to radiotherapy for locoregionally advanced tumours. Recently, the growing understanding of the role of the immune system in tumour suppression has led to the development of immunotherapy for head and neck cancer. The interaction between programmed cell death protein 1, PD-1 and its ligand PD-L1 is one such checkpoint and has recently been a target in systemic therapy efforts. Interaction of these proteins lead to T-cell inactivation and immune tolerance, and increased PD-L1 expression has been found in several subtypes of head and neck cancer (Addeo 2016; De Meulenaere 2017). Inhibitors of this checkpoint have led to recent FDA approval of nivolumab for platinum-refractory recurrent/metastatic head and neck cancer (Harrington 2017).

1.2 HEAD AND NECK CANCER ANATOMY

The volume where squamous cell carcinomas of the head and neck arise is small. It is characterised by short distances between organs and tissues that fulfil important physiologic functions. Both the tumour mass itself as well as the locoregional treatment modalities, surgery, and radiation threaten to compromise these functions because of difficulties to avoid a negative influence on these structures by the treatment. With respect to radiotherapy, the organs that should be spared the influence of treatment are the salivary glands, the parotids, the submandibular glands, and the minor salivary glands in the oral cavity. If the functions of these organs are inactivated, the patient will develop *xerostomia*, a dry mouth, which can be lifelong, with negative effects on oral health and potential complications from teeth and risk of infections, secondary effects on the jaw, problems with swallowing, and nutrition, etc. If the musculature in the posterior part of the pharynx, from below the base of skull caudally to the oesophageal inlet is affected, problems with swallowing will develop *dysphagia*, which can be lifelong, severely complicating food intake and, therefore, daily quality of life. Dysphagia can most probably be accentuated by a concomitant xerostomia. If the temporomandibular joint or the masseter muscle, the structure most responsible for mastication, is affected, *trismus* will appear, i.e. the patient will have difficulty with adequately opening his or her mouth. All these side effects can together boost each other and severely diminish oral health: they can aggravate the effect of radiation on the mandible, reducing the possibilities to heal undue effects on the mandibular bone, and thus increasing the risk of *osteoradionecrosis*. When the larynx is affected by treatment, e.g. by a laryngectomy; this obviously signifies a serious morbidity for the patient. However, also radiotherapy can produce different degrees of *voice changes*, tiredness of voice, and hoarseness. Dose constraints are, therefore, critical. Table 1.1 presents constrains for organs at risk as recommended by the Karolinska University Hospital in Stockholm. Other institutions may have similar but not *exactly* the same dose limits.

TABLE 1.1   Constraints for Organs at Risk for Conventional Radiotherapy

Priority	Organ	Endpoint	Dose
1	Medulla spinalis	Myelitis necrosis	48 Gy, 46 Gy when cytostatics are given, $EQD_{2max} \leq 50$ Gy
2a	Contralateral parotid	Xerostomia	$EQD_{2mean} \leq 20$ Gy
2b	Larynx	Hoarseness, swallowing issues, voice exhaustion	$EQD_{2mean} \leq 40–45$ Gy
2c	Swallowing constrictors	Dysphagia/Swallowing problems	$EQD_{2mean} \leq 45–50$ Gy
2d	Mandible/mandibular joints	Osteoradionecrosis, fracture	$EQD_{2max} \leq 74$ Gy (1 cc)
2e	Contralateral medial half of oral cavity	Acute mucositis, xerostomia	$EQD_{2mean} \leq 30$ Gy
2f	Inner ear	Reduced hearing	$EQD_{2mean} \leq 45$ Gy
3a	Ipsilateral parotid/parotid bilateral	Xerostomia	Minimise/mean=25 Gy
3b	Submandibular gland	Xerostomia	$EQD_{2mean} \leq 24$ Gy
3c	Auditory meatus	External otitis, affected hearing	$EQD_{2mean} \leq 45$ Gy
3d	Trachea	Sensitivity to infections	Minimise, $EQD_{2mean} \leq 44$ Gy

Note: EQD_2 = equivalent dose in 2 Gy fractions, see also the Appendix.

The area (volume) where the tumours can arise begins at the base of the skull and extends to the clavicles. It includes the base of the skull, temporal bone (the external auditory canal, middle and inner ear), paranasal sinuses, nasopharynx, oropharynx (soft palate, tonsil structures, base of tongue, oropharyngeal wall), larynx (supraglottic, glottic, subglottic), hypopharynx (pyriform sinuses, postcricoid area, posterior pharyngeal wall), oral cavity (lips, buccal mucosa, alveolar ridges, floor of mouth, oral tongue, retromolar trigone), major and minor salivary glands, skin, and the neck. Squamous cell carcinomas constitute the majority of histopathologies (about 95%) that may arise in this site, but the various histopathologies, anatomic subsites of involvement, and various sizes of tumour masses result in tremendous discordant variability in the natural course of the diseases for this small anatomic site.

1.3 ANATOMIC LANDMARKS OF SOME SITES AND SUBSITES COMMONLY AFFECTED BY MALIGNANT HEAD AND NECK TUMOURS

Figure 1.1 illustrates the main anatomic landmarks in the head and neck area.

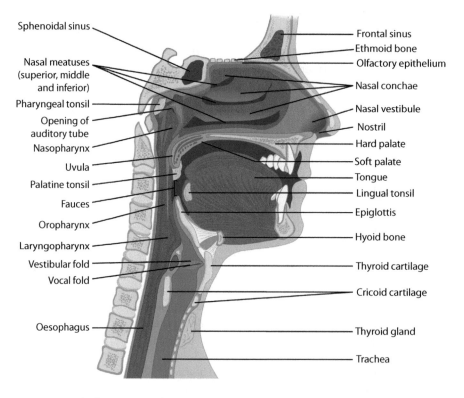

FIGURE 1.1 Sagittal illustration of the head and neck anatomy. (Courtesy of Rice University, Anatomy_of_Nose-Pharynx-Mouth-Larynx, https://opentextbc.ca/anatomyandphysiology/2303 _anatomy_of_nose-pharynx-mouth-larynx/, licensed under a Creative Commons Attribution 4.0 International License, 1999–2016.)

Lips: The lips constitute the facial boundary of the oral cavity. They begin at the junction of the vermilion border with the skin and form the anterior aspect of the oral vestibule. They are composed of the vermilion surface and this portion of the lip comes in contact with the opposing lip. The upper lip meets the structures of the nose in the so-called sulcus nasolabialis. The lower lip is restricted caudally against the chin by sulcus mentolabialis. The lips are made up of musculature, the oral orbicular muscles.

Cheeks (buccal mucosa): The cheeks are mostly made up of musculature, the buccinator muscles. The cheeks are covered by mucosal surfaces, extending from the line of contact of the opposing lip to the pterygomandibular raphe posteriorly. A body of fat, corpus adiposum buccae, runs posteriorly between the masseter muscle and ramus mandibulae. In the most ventral parts of the submucosa, there are submucous salivary glands.

Palate: The palate consists of two parts, the hard palate, palatum durum, constituting the ventral two-thirds of the roof of the oral cavity and the soft palate, palatum molle, constituting the most posterior part of the roof of the oral cavity. The mucosal lining of the hard palate is gently fixed to the underlying bone while it is much looser in the soft palate. The soft palate constitutes the oral part of velum palatinum, which is made up of robust fibrous tissue and contains musculature, vessels, and nerves of great importance for the process of chewing and swallowing. The minor salivary glands of the oral cavity are most frequent in the palate.

Floor of mouth: The floor of mouth is a space extending from the lower alveolar ridge to the undersurface of the tongue. It overlies the mylohyoid and hyoglossus muscles. The mylohyoid muscles extend from linea mylohyoidea on the inside of the mandible and meet in the midline in a fibrous band, covering a distance from the inside of the chin to the hyoid bone. The floor of mouth is reinforced by two other muscles, posteriorly by the geniohyoid muscle and ventrally by the anterior part of the digastric muscle. The muscular part of the floor of mouth is not complete since mylohyoid muscle does not reach the posterior ramus of the jaw.

Tongue: The tongue is a mobile sort of muscular tissue, which can be divided into two parts. The most mobile part, corpus linguae, is situated in the oral cavity while the rest of the tongue, the base of tongue or radix linguae, is situated in the oropharynx. This latter part of the tongue belongs together with tissues in the soft palate and in the tonsils to the Waldeyer's ring. The anterior two-thirds of the tongue, the oral tongue, extends from the circumvallate papillae to the undersurface of the tongue at the junction of the floor of mouth. A fibrous septum divides the tongue into right and left halves. The oral tongue is commonly demarcated into four anatomic areas: (1) the tip, (2) lateral borders, (3) dorsal surface, and (4) undersurface (ventral surface). There are six pairs of muscles that form the oral tongue: three of these are extrinsic and the rest are intrinsic. The external muscles are m. genioglossus, m. hyoglossus and m. styloglossus, and are those muscles that can move the tongue in the oral cavity, prevent it from falling backwards, etc. The intrinsic muscles are situated in the depth of the tongue and include the lingual, vertical, and transverse muscles. These muscles can alter the form of the tongue during speech and swallowing. The base of the tongue, belonging to oropharynx is bordered ventrally by the so-called sulcus terminalis. Caudally the base of tongue is bounded by the most cranial part of the supraglottic larynx, i.e. the epiglottis. The base of tongue is characteristically composed by an abundance of lymphoid tissue, constituting together with the tonsils and the soft palate what is called the Ring of Waldeyer.

Tonsils: The tonsils belong to the oropharynx which is situated behind the oral cavity, has a wide dorsal wall and two tiny lateral borders. Caudally, it continues into the hypopharynx. The tonsils are situated laterally in the oropharynx with three main parts: (1) the tonsilla fossa, (2) the anterior, and (3) posterior pillars. Ventrally the tonsils abut on the retromolar trigone, which is a part of the oral cavity, posteriorly the dorsal wall of the oropharynx, cranially by the roof of the tonsillar fossa, and caudally by the epiglottis.

1.4. INCIDENCE OF HEAD AND NECK CANCER

Over 500,000 new cases of head and neck squamous cell carcinoma are reported annually worldwide with 40,000 new cases reported in the United States with 7890 deaths (Siegel 2015; Torre 2015). These tumours represent about 2.8% of all new cancer cases in the United States and can, therefore, be said to be rare. In Sweden, as a representative of the Nordic countries, squamous cell carcinomas of the head and neck are likewise not very common, representing 2.0% of all new cancers. In the United States, it was estimated that 55,070 new cases would occur in 2014 with about 12,000 patients' dead from their disease. In Europe, it was estimated in 2010 about 139,000 new cases of head and neck cancer per year. The same year it was reported a 1-year survival rate of 72%, while the 5-year survival rate was only 42% for head and neck cancer in adults in a European study (Gregoire 2010). The tumours can arise from several subsites in the area, within the oral cavity, oropharynx, larynx, hypopharynx, nasopharynx, and the paranasal sinuses. Tobacco and alcohol continue to remain the two major risk factors for head and neck cancer. Incidence rates tend to vary widely around the world and even within populations. The black population in the United States tends to have a higher incidence rate than whites and Hispanics. On most continents, the rates tend to be higher in males than in females with a ratio from 2:1 up to 5:1 (Parkin 2002). However, over the last decade, there has been a shift in the primary site distribution of all head and neck tumours with a steady increase in oropharyngeal carcinomas and a decline in tumours arising in other sites, mainly the larynx and hypopharynx. This change has also been observed in parallel with the identification of exposure to high-risk oncogenic HPV as a risk factor for developing oropharyngeal cancer. During the same period, a decline in cigarette smoking has been recorded in many countries (Sturgis 2007; Gillison 2012). The incidence of HPV+ oropharyngeal cancer has been reported to increase by 225% in the United States between the years of 1988 and 2004, while HPV–cancers declined by 50%. In North America, 56% of oropharyngeal cancers are reported to be HPV+, followed by 52% in Japan, 45% in Australia, 39% in northern and western

Europe, and 13% in the rest of the world (Chaturvedi 2013; Gillison 2014). For instance, in Sweden, there was a significant increase in the incidence of base of tongue cancer from around 0.3/100,000 person-years between 1970 and 1999 to around 0.5/100,000 person-years in 2000–2007. In addition, there was an increase in the prevalence of HPV+ base of tongue carcinomas in Stockholm County from 58% during 1998–2001 to 84% in 2006–2007. It is, therefore, suggested that HPV infection is responsible for the increased incidence of base of tongue cancer in Sweden (Attner 2010). Almost all patients in Stockholm County diagnosed with tonsillar squamous cell carcinoma have HPV+ tumours: from the 1970s to 2000–2007 there was a continuous increase in the proportion of HPV+ tonsillar cancer from 23% to 93%. There was almost a doubling in the age-standardised incidence of HPV+ tumours per decade; although the incidence of HPV– tumours instead initially showed an increase from the 1970s to the 1980s and then declined in the 1990s and onwards. The decrease of HPV– tonsillar cancer is, therefore, similar to that of lung cancer and follows the trend of other smoking-related cancers in Sweden (Näsman 2009).

Patients with HPV-associated head and neck cancer tend to be younger, typically middle-aged, non-smoking men of higher socioeconomic status, and with a history of exposure to multiple sexual partners. Some of these patients may have a history of previous smoking but most are not current smokers. Analyses from clinical trials indicate that patients with locoregionally advanced HPV+ tumours experience improved response to treatment and overall survival and progression-free survival compared with HPV– cancers. A systematic review including prospective or retrospective studies showed that patients with p16-positive oropharyngeal cancer had a better prognosis and fewer rates of adverse events compared with those with p16-negative disease (Wang 2015). This change in panorama has now dramatically reformed research directions within the head and neck cancer academic communities in many countries. There is an increasing tendency now to recognise squamous cell carcinoma of the head and neck as two separate entities, i.e. whether the disease is HPV+ or HPV–. The distinction is driving research concerning biology, mutational landscape, predictors of response to therapy and survival outcome. While de-escalation has been one focus of treatment development for HPV+ cancer patients in an effort to decrease late complications, more intensified options, including new targeted agents, has been one approach for the HPV– patients in order to improve the poor overall survival of this group.

1.5 RISK FACTORS AND PATHOGENESIS OF HEAD AND NECK CANCER

1.5.1 Tobacco and Alcohol Consumption

Tobacco and *alcohol* continue to remain two important risk factors for head and neck cancer with their carcinogenic risks summarised in several working group reports by the International Agency for Research on Cancer (IARC 1986, 1988). Cumulative evidence easily fulfils the criteria for causality between cigarette smoking and the development of head and neck cancer (Franceschi 1990; Talamini 2012). The demonstration that the risk of developing head and neck cancer rises with increasing numbers of cigarettes smoked per day supports such a causal relationship. Current smokers have approximately a 20-fold higher risk of oropharyngeal and laryngeal cancers than lifelong non-smokers. The increased relative risk of developing head and neck cancer in the heaviest smokers is quoted as 20 to 40 times that of nonsmokers. In addition to the duration of smoking and number of cigarettes, the type of cigarettes and the age, sex, and race of the smoker also influence the relative risk. Despite limited data there seems to be an association between cigar and pipe smoking and an increased risk of head and neck cancer even if the risk appears lower, between 1.9 and 10.3 times that of nonsmokers. Exposure to environmental smoke has also been implicated in several publications as a risk factor; individuals who were exposed to the highest levels of environmental cigarette smoke were up to four times more likely to develop head and neck cancer (Zhang 2000). Also, tobacco consumed in smokeless forms, as is common in many cultures around the world like betel quid in certain parts of Asia, increases the risk of head and neck cancer with a preference in the oral cavity, the tongue, floor of mouth or buccal mucosa.

Alcohol consumption is also known as an independent risk factor for head and neck cancer. For those individuals who drink heavily but do not smoke the increased relative risk ranges from 5.0 to 11.6 with an increased significance with higher number of drinks consumed (Franceschi 2000; Goldstein 2010). When alcohol and tobacco are consumed together the risk increases multiplicatively rather than additively (Blot 1988; Maier 1992). Very high odds ratios have been reported for heavy consumers of both alcohol and tobacco (Franceschi 1999; Tarvainen 2017).

1.5.2 The Human Papillomavirus

Since the early 1990s, *HPV* has also been suggested as a risk factor for head and neck cancer and, as mentioned before, has been an important reason for a changing incidence panorama. Numerous studies have shown a high prevalence, 45% to 75% of HPV in tonsillar cancer tissue (Snijders 1992; Gillison 2000). However, the presence of HPV in tumour tissue does not necessarily imply viral involvement in the carcinogenesis but may reflect a transient infection. An important epidemiologic finding was when Mork and coworkers in a case-control study showed that sign of an earlier HPV infection, as measured in sera, was an independent risk factor for the development of tonsillar or base of tongue cancer. This finding was important since it showed that HPV infection preceded the development of cancer by on average 9.4 years. The odds ratio for developing oropharyngeal cancer was 10.2 (95% CI 2.4–42.9) and base of tongue cancer 20.7 (95% CI 2.7–160.1) (Mork 2001). Other epidemiologic reports show links between oral and oropharyngeal cancer. HPV infection has been found to be associated with an increased risk for oral and oropharyngeal cancer. Risk estimates were stronger when restricted to oropharyngeal cancer (Smith 2004; Hansson 2005). When using DNA as well as RNA in situ hybridisation, the viral genome and its transcription products performed on HPV16 have been located in cancer cells, both at the primary site and in the metastasis, but not in the surrounding stroma. HPV16 E6 and E7 mRNA expression is known to be essential for transformation in cervical cancer; these oncogenes have also been found in oropharyngeal cancer, thus also suggestive of a causal role for HPV (Wiest 2002). Several studies, in line with cervical cancer, have found that sexual behaviour, such as a high number of sexual partners, is a high risk for cancer of the oropharynx (Schwartz 1998; Rosenquist 2005). Also, certain sexual behaviour has been reported to be strongly associated with HPV+ tonsillar cancer including a history of performing oral sex and oral-anal contact (Ritchie 2003; Smith 2004). Individuals at increased risk of tonsillar cancer include those with a history of HPV-associated anogenital malignancy, women over the age of 50 years with a history of in situ cervical cancer, and husbands of women with in situ or invasive cervical cancer (Frisch 1999; Hemminki 2000). It has also been shown that patients with Fanconi anaemia have increased risk of head and neck cancer with as much as 500 to 700-fold. This may be attributable to an increased genetic susceptibility to HPV-mediated tumour genesis (Kutler 2003; Lowry 2003).

1.5.3 Other Risk Factors

There are also other risk factors for head and neck cancer. These include genetic predisposition, previous head and neck cancer, history of malignant diseases in the immediate family members, exposure to ionising radiation, nutritional disorders or habits, vitamin deficiencies, iron-deficiency anaemia, poor oral hygiene, chronic infections, and long use of badly fitting prostheses. It has been implied that head and neck cancer patients have increased chromosomal sensitivity to carcinogen exposure that predisposes them to developing cancer (Cloos 1996; Schantz 1997). Laboratory in vitro studies have shown that cells from head and neck cancer patients suffer more chromosome breaks upon exposure to a mutagen than do normal control cells (Spitz 1989). Moreover, young patients with these tumours who were nonsmokers demonstrated a mucosa that was particularly mutagen sensitive (Schantz 1989).

Immunosuppression may predispose individuals to an increased risk of head and neck cancer. In patients with a renal transplant, there is increased risk of lip carcinomas and an increased risk of oral cavity carcinomas have been reported in patients with HIV but also in patients with cardiothoracic transplants (Pollard 2000). These tumours are in general much more aggressive with a decreased overall survival (Preciado 2002).

1.6 PATHOLOGY AND PATHOGENESIS

Most cancers in the head and neck area are squamous cell carcinomas or one of their variants, lymphoepithelioma, spindle cell carcinoma, verrucous carcinoma, and undifferentiated carcinoma. In addition, there are adenocarcinoma, salivary gland tumours, melanomas, Kaposi sarcomas, other sarcomas, lymphomas, and solitary plasmocytomas. Merkel cell carcinoma usually arises from the skin but can also be seen in mucous membranes. A molecular tumour progression model was originally described by Fearon and Vogelstein (Fearon 1990). According to this model tumours progress by activating oncogenes and by silencing tumour suppressor genes. Both of these processes produce a growth advantage for a clonal population of cells. The model also states that specific events occur in a distinct order, a so-called multistep carcinogenesis that is not necessarily the same for all tumours. Califano and coworkers proposed a tumour progression model for head and neck cancers using allelic imbalance as a molecular marker for oncogenic amplification or inactivation of tumour suppressor genes. They identified p16, p53 and Rb as candidate tumour suppressor genes, and cyclin D1 as a candidate proto-oncogene (Califano 1996). This work supports the hypothesis that clonal genetic changes occur early in the histo-pathologic continuum of tumour progression. In histologically benign squamous hyper-plasia one can already identify clonal populations of cells that share genetic abnormalities with invasive head and neck cancer. Various mucosal lesions are frequently encountered in the oral cavity and it is well known that some of these are burdened with risk of definitive malignant transformation but the incidence of tumour progression varies considerably. Different criteria have been used for prediction of cancer development such as macroscopic appearance of the lesion or the microscopic grading of dysplasia. However, the macroscopic appearance of lesions described as erythroplakia or leukoplakia is not reliable for cancer prediction and both macroscopic and microscopic evaluations of preneoplastic lesions show great intra- and interobserver variations. Invasive squamous cell cancer of the larynx can develop from progressive dysplasia and cancer in situ (CIS). Histopathological evaluation of CIS lesions is difficult, shows poor reproducibility and is an unreliable predictor of risk for progression to invasive cancer. The introduction of newer molecular assay techniques has greatly increased the ability to detect important genetic changes and thereby better understand relevant cancer biology: comparative genomic hybridisation, in situ hybridisation, single-nucleotide polymorphism, and microarray technology.

Head and neck cancer caused by exposure to smoking and alcohol is more likely to be associated with mutations in tumour suppressor genes and specifically TP53. This also makes these cancers less sensitive to combined treatments with radiotherapy and cytostatic therapy. The Cancer Genome Atlas Network used whole-genome sequencing on tumour tissue from 279 patients with head and neck cancer and reported data confirming a nearly universal loss of function of TP53 mutations and the CDKN2A gene inactivation in high frequency in HPV− cancers. In HPV+ head and neck tumours recurrent deletions and mutations with tumour necrosis factor receptor-associated factor 3, PIK3CA and amplification of cell cycle gene E2F1 were identified (Cancer Genome Atlas Network 2015).

Human papillomavirus is a small DNA virus with predilection to cutaneous or mucosal squamous epithelium located in a subset of head and neck cancers mostly in the oropharynx, but also in the cervix and the anogenital region. The oncogenic types HPV 16, HPV 18, HPV 31 and HPV 33 are sexually transmitted and are those considered high-risk factors for malignant transformation. In most individuals, the infection clears. However, in those individuals where the immune response is not strong enough, viral DNA is integrated in the host genome and the virus prefers to target the highly specialised epithelium that lines the tonsillar crypts (Kim 2007). Replication takes place within the infected cell nucleus and is dependent on S-phase entry (Deng 2004). Once the virus integrates its DNA genome within the host cell nucleus, expression of the oncoproteins E6 and E7 takes place. The E6 protein degrades p53 through ubiquitin-mediated proteolysis, leading to substantial loss of p53 activity and the oncoprotein E7 binds and degrades the retinoblastoma pocket protein Rb, resulting in cell-cycle disruption, proliferation, and malignant transformation. The functional inactivation of Rb results in the overexpression of p16 INK4A encoded by CDKN2A in the HPV-infected tissue. HPV+ tumours are consequently characterised by high levels of p16 INK4A (Nevins 2001). Several studies have also shown a very high correlation between high levels of p16 INK4A and HPV positivity and it has been suggested as a clinically useful surrogate marker for HPV. Immunohistochemical (IHC) analysis of the tumour tissue for p16 INK4A is now used in the head and neck community as the initial test of choice and a surrogate marker to identify high-risk HPV infection in tumour tissue.

CHAPTER 2

Clinical Aspects of Head and Neck Cancer

2.1 INTRODUCTION

According to the latest Globocan report released by the International Agency for Research on Cancer, the estimated incidence of head and neck cancer (HNC) in 2012 among men represents 7% of all cancers, while this is 2.6% in women (Ferlay 2012). HNCs comprise tumours arising from the following anatomic locations: oral cavity, pharynx (nasopharynx, oropharynx, hypopharynx), larynx, paranasal sinuses, nasal cavity and salivary glands. The incidence of HNCs caused by known risk factors such as tobacco smoking and alcohol consumption has decreased lately (Chaturvedi 2008). Nevertheless, a relatively new entity of head and neck squamous cell carcinoma, which is mostly located in the oropharynx was shown to be on the rise and was associated to the human papillomavirus (HPV) (Ang & Sturgis 2012). This HNC represents a distinct entity from the non-HPV tumours, holds different biological characteristics and responds differently to treatment (Ang & Sturgis 2012; Urban, Corry & Rischin 2014).

While early-stage head and neck tumours are easier to handle, treatment of advanced cancers encounters a higher degree of difficulty given by various factors. Beside the patient's medical history, radiobiological factors such as hypoxia, proliferative ability, cancer stem cell phenotype and intrinsic radioresistance are strong influencing factors concerning the optimal treatment choice. Predictive assays for oxygen status, proliferation and radioresistance have been previously designed in vitro with the aim to be routinely used in the clinics (Hall 2000). However, these assays created more challenges than solutions (Table 2.1) thus today we rely on modern imaging techniques to tackle these problems and to guide treatment.

The main purpose of predictive assays would have been to distinguish between tumours with distinct radiobiological features in order to design the treatment according to the tumours' response. Decades of experience in head and neck radiotherapy have led to the

TABLE 2.1   Predictive Assays for Tumour Response to Radiotherapy and Their Limitations

Predictive Assay	Oxygenation Status	Proliferative Potential	Intrinsic Radioresistance
Purpose	To identify the patient group that would benefit from hypoxic cell sensitisers	To differentiate between tumours with slow and fast proliferation	To correlate cell line radiosensitivity with tumour response to radiation
Technique	Polarographic needle electrode Endogenous/exogenous markers	Kinetic parameter measurements: length of S phase, potential doubling time; labelling index.	Dose-response curves; Colony growth (MTT), micronucleus, chromosomal, DNA damage (Comet) assays.
Limitation	Invasive; Unreliable (biopsies); Costly and time consuming; Require high level expertise.	No robust correlation between kinetic parameters and treatment outcome; Time consuming.	Highly time consuming.
Present/Future	Hypoxia-specific PET radiotracers: F-MISO; F-FAZA; Cu-ATSM; other radiotracers BOLD (blood oxygen level-dependent) MRI	Proliferation-specific PET radiotracers: F-FLT; F-ISO-1; ^{11}C-based radiotracers.	HPV-status based identification of more radioresponsive tumours.

conclusion that these cancers are usually hypoxic and more rapidly proliferating than other malignant tumours. Therefore, treatments have been adjusted according to these radiobiological properties. For instance, the change from conventionally fractionated radiotherapy to altered fractionation schedules was determined by the effect of various accelerated repopulation mechanisms on tumour behaviour during therapy. Thus, accelerated radiotherapy and even so accelerated-hyperfractionated radiotherapy, have improved both loco-regional control and overall survival in patients with unresectable head and neck tumours (Overgaard 2003; Bourhis 2006; Saunders 2010). The development of bioreductive drugs (hypoxia-activated prodrugs) as hypoxic cell sensitisers have opened new therapeutic avenues for selected groups of patients.

However, normal tissue toxicity is often a limiting factor in head and neck radiotherapy. The need for organ preservation in advanced HNCs has led to the implementation of novel techniques such as 3D-conformal radiotherapy (3D-CRT), intensity modulated radiotherapy (IMRT), image-guided radiotherapy (IGRT) and volumetric modulated arc therapy (VMAT), which deliver a more conformal dose to the target thus allowing for better normal tissue protection. Research in the field of radioprotectors and radiomitigators to assist in the sparing of normal tissue has shown improved results in the radiotherapy of selected head and neck patients. The future of HNC probably resides in targeted therapies and (radio)immunotherapy, although currently these approaches fail to keep pace when compared to the management of other solid malignancies.

When applied to combined modality treatment, radiobiological principles have been shown to improve the outcome of HNC radiotherapy. As mentioned above, alteration of the conventional fractionation and treatment acceleration have led to the optimisation of the time factor in treatment scheduling, which in turn has enhanced tumour response (Peters 1997). What is expected from future research is the identification and exploitation of head and neck tumour-specific antigens and families of tumour antigen to increase tumour responsiveness with no added normal tissue toxicity.

Early surveillance and modern imaging techniques have greatly contributed to the improved treatment results over the last few decades. The critical role of positron emission tomography (PET)/computed tomography in the management of HNC has been established in the early 1990s, when independent research groups have employed FDG-PET (flourodeoxyglucose for PET) for tumour detection and prediction of treatment outcome (Bailet 1992; Seifert 1992; Greven 1994). Next to FDG, hypoxia-specific radiotracers have been successful in personalised treatment design. Therefore, pretreatment ^{18}F-MISO (fluoromisonidazole) uptake has been shown to be an independent prognostic indicator following treatment of HNC (Rajendran 2006). ^{18}F-FAZA (fluoroazomycin-arabinoside) is another radiotracer that has been intensively studied relative to hypoxia and successfully implemented into clinical settings (Souvatzoglou 2007). The proliferative ability of HNC can also impact on tumour response to treatment. To tackle this aspect, ^{18}F-FLT (fluorothymidine) has been trialled by several groups demonstrating its effectiveness in accurate staging of HNC (Vojtisek 2015) and as an early predictor of outcome during chemo- and radiotherapy (Hoeben 2013). PET-based imaging has become a critical component of the treatment planning in HNC, being a valuable treatment guide during the course of chemoradiotherapy and an important player in personalised patient management.

2.2 SHORT HISTORICAL PERSPECTIVE OF HEAD AND NECK CANCER TREATMENT

The first evidence of human cancer in the head and neck area has been found in Egyptian skulls from the First Dynasty (3400 BC) (Ackerknecht 1958). The Egyptian papyri describe the tumours as 'swellings' and 'ulcers' that were mainly treated by excision. The first to recognise the differences between benign and malignant tumours was Hippocrates, considered the father of medicine. In the 5th century BC Hippocrates already described the nasopharyngeal fibroma, a benign tumour that was named as such only in 1940.

While Greek and Roman medical writings were the foundation of modern medicine, there are only few mentions of specific cancer treatment techniques during those times. Claudius Galenus (also known as Galen, 130 AD) a Greek physician and surgeon, has further developed Hippocrates' practice and authored a special book on tumours, where he described both benign and malignant lesions, including their growth pattern that was 'outside of nature'. Until the employment of anaesthesia and surgical excision in the 11th century no significant progress was made in the management of HNC.

The first link between tobacco smoking and cancer was established in the 18th century by the London physician John Hill (Redmond 1970). He suggested this association in his work 'Cautions against the immoderate use of snuff' published in 1761, where he reports six cases of tobacco-related lesions that he names 'polypusses'.

A pivotal change in the understanding and management of cancer came with the employment of the microscope in pathological anatomy. Consequently, studies on biopsies have created new avenues to bring insights into the development of cancer. The epithelial origins of certain cancers have been demonstrated by Holland (1843) and Remak (1852) (Ackerknecht 1958).

Nevertheless, significant achievements in tumour control occurred after the discovery of X-rays and radium. Brachytherapy, using radium surface moulds and plaques, has been practiced since 1903 (Mould 1993), while for head and neck carcinomas surface moulds have been first used in the 1920s. Deep-seated HNCs were managed with interstitial brachytherapy. The first interstitial treatment technique with radium was reported by the German physician H. Strebel (Mould 1993). This technique was developed to avoid the side-effects caused by X-ray treatments which affected much of the healthy tissue in order to deliver the tumoricidal doses to the target volume. Interstitial treatment with radium salt was shortly replaced by radon gas-filled glass seeds, as radon was cheaper and due to its short decay time the seeds could be left in place. Brachytherapy treatments became more targeted and safer during the second half of the 20th century when afterloading applicators have replaced manual loading. Nowadays, high dose rate (HDR) brachytherapy using iridium 192 is widely available and used for both intracavitary and interstitial brachytherapy.

Brachytherapy was followed by the introduction of multimodality treatment due to improved understanding of the molecular changes underlying cancer growth. Since the early days, surgery was the first treatment choice for operable HNC. This has not changed in present times. Nevertheless, radiotherapy and chemotherapy, either alone or in combination were used with different degrees of success in the treatment of unresectable and locally advanced carcinomas. Today, concurrent chemoradiotherapy is the standard of care in head and neck oncology for locoregionally advanced cancers (Pignon 2009).

2.3 OVERVIEW OF TUMOUR SITE-RELATED TREATMENT TECHNIQUES FOR THE MANAGEMENT OF HEAD AND NECK CANCER

Although HNCs are usually mentioned as a tumour group, the common anatomical sites where squamous cell carcinomas develop exhibit different radiobiological properties which require different treatment approaches. The following sections discuss the site-related treatment techniques from a historical perspective, while focusing on the modern treatment methods for HNC.

2.3.1 Oral Cancers

The first identification of oral cancers dates back as early as 460 BC. Hippocrates attributed a malignant ulcer at the edge of the tongue to chronic mechanical trauma induced by sharp teeth rubbing against the tongue (Thumfart 1978). The Roman encyclopaedist Celsus was among the first ones to report radical operations of the lip (Ackerknecht 1958). The first to remove a lingual cancer was the Italian physician Marchetti in the early 1700s. For this procedure he has used cautery, which seemed to be the treatment choice until the 1800s when the knife started to be used for such operations (McGurk 2000). Removal of tongue cancers has been practiced during the 18th century well before the implementation of anaesthesia. One type of procedure consisted of tumour removal through tissue strangulation by encircling with a line of sutures (McGurk 2000) which later led to the development of ecraseurs.

Starting with the late 1800s when the medical world was fascinated by the study of biopsies, the approach towards HNC became more refined. In the early 1900s, Albert Broders, a surgical pathologist from the Mayo Clinic has defined a grading system for the squamous cell carcinoma of the lip, while also applying the method to other cancers of the head and neck (Broders 1920).

In modern times, surgery through neck dissection and/or radiotherapy is customarily one of the treatment choices for oral cancers. To increase tumour control when treating with curative intent but also for palliation, chemotherapy has been employed in combination with the existent approaches. A retrospective study has indicated that taxane-based regimens in the form of neoadjuvant chemotherapy might play a role in the treatment of oral cancers (Sturgis 2005). Confirming this observation, pre-clinical studies on nude mice examining the effect of induction chemotherapy on overall treatment outcome have shown an inhibitory effect of neoadjuvant chemotherapy on cervical lymph node metastasis of oral squamous cell carcinoma (Kawashiri 2009). Given the fact that about half of oral squamous cell carcinoma present with cervical lymph node metastasis, Sato et al. have investigated the role of an oral fluorouracil compound, S-1, looking at the inhibitive effect of the drug on metastatic development. The metastatic suppression induced by the agent was proven in nearly half of the tested mice, the results being indicative of the important role of neoadjuvant chemotherapy in the management of oral cancers (Sato 2009). Chemoradiotherapy for oral cancers commonly leads to serious side-effects such as xerostomia, mucositis and even osteoradionecrosis. To overcome these unwanted tissue toxicities, other treatment methods have been proposed and trialled.

A less toxic approach towards oral cancers implemented in the early 1990s is photodynamic therapy (PDT) also known as photoradiation therapy. PDT has been used in patients with field cancerisation of the oral cavity, as an obvious choice over highly disfiguring surgery (Grant 1993). Photoradiation therapy involves a photosensitiser in the form of a tumour-specific photoactive dye (usually 5-aminolevulinic acid or photofrin), which is activated by light exposure of a particular wavelength (typically with 50–100 J/cm^2 red or violet laser light) in the presence of oxygen. The photosensitiser selectively accumulates in the abnormal cells and after light-activation reacts with the available tissue oxygen. This process consequently leads to cell damage and local tissue necrosis (Nauta 1996). PDT is more suitable for superficial lesion due to limited penetration of the clinically used agents (Fan 1996). Alongside non-thermal lasers, light-emitting diodes (LEDs) have been employed in PDT (Kvaal 2007). Recent reviews of the literature on PDT confirm the efficacy of this method on oral cancers and precancerous lesions and encourage more research for the establishment of the optimal treatment protocol (Prazmo 2016, Saini 2016).

There is also a small incidence of HPV+ cases (5.9% of all oral cancers) among patients with oral squamous cell carcinomas (Lingen 2013). It is known that HPV+ HNCs usually have a better response to treatment than those that tested negative for HPV. Nevertheless, it is not completely elucidated whether all HPV+ HNC should be treated differently from their non-HPV counterparts or only the oropharyngeal carcinomas.

2.3.2 Salivary Gland Cancer

Salivary gland cancers occur mainly in the parotid gland, however, predisposed individuals are also at risk to develop tumours of the submandibular, sublingual and the minor salivary glands. One of the first and most comprehensive reports on salivary gland cancers has been published by H. Ahlbom, a Swedish radiotherapist who has dedicated part of his report to tumour pathogenesis, while trying to elucidate the lack of resemblance with the tissue of origin and taking the first steps to categorise salivary gland tumours (Ahlbom 1935).

The most common treatment approach for resectable salivary gland cancers is surgical excision, with or without postoperative radiotherapy. Unresectable tumours are treated with concurrent chemoradiotherapy using either cisplatin- or taxol-based cocktails.

As an alternative to surgery, radiotherapy as a single modality treatment has also been reported to lead to significant long-term benefits (Chen 2006). However, one of the most successful therapies for salivary gland cancers is fast neutron therapy, which offers high locoregional control and survival rates, being recommended as initial primary treatment (Douglas 1996). A report on 279 patients treated with curative intent for salivary gland cancer has been published by the University of Washington Cancer Centre team (Douglas 2003). The 6-year actuarial overall survival and locoregional control was 59%.

A combined Radiation Therapy Oncology Group (RTOG) randomised clinical trial of neutrons versus photons for non-resectable or recurrent salivary gland tumours accrued 25 patients, with 13 patients in the neutron arm and 12 in the photon arm (Laramore 1993). The neutron arm has received 12 fractions of neutrons to 16.5 to 22 Gy while the photon arm was delivered 70 Gy for 37 fractions or 55 Gy for 20 fractions. The superiority of fast neutron therapy at 2 years was obvious, with local control and survival of 67% and 62% for neutrons versus 17% and 25% for photons.

Despite the superior results obtained with neutrons, the role of fast neutron therapy in the management of HNCs is limited to salivary gland tumours (Laramore 2009).

2.3.3 Cancer of Nasal Cavity and Paranasal Sinuses

According to some historians, cancers of the nasal cavity were common even in prehistoric times due to wood smoke inhalation in scantily ventilated sheds. The Arab Rhazes (900 AD) has been able to differentiate between nasal polyps and cancer (Ackerknecht 1958).

Nasal and sinus cancers are rare compared to other types of head and neck tumours. The primary treatment choice is usually surgical removal of the tumour. Radiotherapy can result in similar outcome when administered to the early stage disease. However, advanced cases should be managed with both surgery and postoperative radiotherapy in order to improve local control and survival, and to reduce normal tissue complications (Katz 2002).

Modern treatment techniques such as 3D-CRT and intensity-modulated radiotherapy have been shown to minimise the occurrence of late complications that are common with conventional radiotherapy (Hoppe 2007). Helical tomotherapy combined with chemotherapy for locally advanced squamous cell carcinoma of the nasal cavity and paranasal sinus has been proven to be efficient in both tumour control (3-year overall survival of 59.2%; local control 80.2%) and orbital organ preservation (77.8% after 3 years follow-up)

(Chen 2016). Multimodality treatment that employs modern radiotherapy techniques provides better outcome thus better quality of life.

2.3.4 Nasopharyngeal Carcinoma

Nasopharyngeal carcinoma arises in the epithelial lining of the nasopharynx. Treatment results of nasopharyngeal cancers are often reported separately from other head and neck carcinomas due to differences in histopathology, nodal stage, incidence of metastases and overall response to therapy. There is a strong geographical variation related to the incidence of nasopharyngeal cancer, the Asian population (east and southeast) being the most predisposed. The most common risk factor associated to this neoplasm is the Epstein-Barr virus infection (Lo 2012).

Operable nasopharyngeal carcinomas, either primary or recurrent, can be managed by surgical resection. In order to achieve radical removal of the tumour, infratemporal fossa surgery is employed. Despite its complexity, this procedure is common in several specialised clinical centres (Fisch 1983). Over decades, numerous changes in surgical procedures of skull-based carcinomas have occurred due to better understanding of the surgical anatomy, thus better management of the surrounding critical structures.

Nasopharyngeal carcinoma is probably the most radio- and chemo-responsive among all squamous cell carcinomas of the head and neck. Consequently, for early-stage disease, radiotherapy is the primary curative treatment, leading to very high locoregional control (Fu 1998). Nowadays, IMRT is the preferred radiation technique, given the suitable sparing of the normal tissue as compared to 2D radiotherapy. Clinical studies suggest that the recent improvement in locoregional control and survival in patients treated for nasopharyngeal cancer is greatly attributed to IMRT (Chua 2016). As with other HNCs, combined chemoradiotherapy leads to better results for late stages of the disease than radiotherapy alone. Cisplatin-based treatments often combined with 5FU are the most common choices of chemotherapeutic agents. The use of new therapeutic targets such as the epidermal growth factor receptor (EGFR) and the vascular endothelial growth factor (VEGF) offer promising results in patients with advanced cancers that are at risk in developing distant metastases (Dorsey 2013; Bossi 2016).

2.3.5 Oropharyngeal Carcinoma

Similar to other HNCs, oropharyngeal carcinoma has been attributed to risk factors such as smoking and alcohol consumption. However, a new entity has been increasing in incidence among oropharyngeal cancer patients, and that is related to HPV infection. Cancers associated to HPV have been shown to be more responsive to therapy than those triggered by smoking and drinking.

Most commonly, multimodality treatment is employed in the management of oropharyngeal carcinomas whether resectable or unresectable. A multimodal intensification regimen consisting of perioperative cisplatin-based chemotherapy, surgery and postoperative chemoradiotherapy were shown to lead to excellent disease control rates (91% locoregional control) and long-term survival (73%) in patients with advanced, resectable cancers (Schuller 2007).

While smoking-related advanced stage oropharyngeal cancers require more aggressive treatment (i.e. intensification regimen), the trend in HPV-associated oropharyngeal cancers is towards dose de-intensification, due to their better treatment response. Recent studies have shown that while leading to the same tumour control, dose de-escalation decreases treatment-related morbidity. Some promising dose de-escalation strategies include the reduction of overall radiation dose, elimination of chemosensitising agents, replacement of chemotherapy with more targeted agents (monoclonal antibodies), and the employment of less invasive surgery (Kelly 2016).

2.3.6 Hypopharyngeal Carcinoma

With the availability of external radiotherapy, the hypopharynx was one of the first deep-seated tumours treated with a fractionated radiotherapy regimen by Coutard in 1919 (Fletcher 1986).

Cancer of the hypopharynx is an aggressive neoplasm that is associated with poor prognosis, which therefore requires an aggressive therapy. As with other head and neck malignancies, surgery and radiotherapy remain the first treatment choice for early stage disease, though patients tend to present with locally advanced disease. For locoregionally advanced tumours, the choice of treatment is dictated by numerous factors such as: performance status, invasion pattern, laryngeal involvement and the presence of distant metastasis. Surgery is still an indispensable part of treatment in advanced hypopharyngeal cancer, and is commonly combined with chemoradiotherapy. Administered prior to surgery or radiotherapy, chemotherapy was shown to help in organ preservation (Shirinian 1994). Modern treatment techniques, such as IMRT and altered fractionation regimens used in combination with cisplatin-based chemotherapy also improve treatment outcome (Paximadis 2012). Nevertheless, to date, there is no consensus regarding the standard treatment approach in advanced hypopharyngeal carcinomas (i.e. surgery followed by adjuvant radio/chemoradiotherapy or definitive chemoradiotherapy) (Harris 2015).

2.3.7 Laryngeal Carcinoma

Considerable knowledge of the anatomy of the larynx originates from the Byzantine Empire through the great physician Oribasius (325–403 AD) who has described the function of the vocal cords in his work 'On the larynx and epiglottis' (Assimakopoulos 2003). Byzantine texts also describe operations of the larynx, though the lack of appropriate instrumentation would raise questions concerning the efficiency of such interventions. Much later, Giovanni Morgagni, an Italian anatomist from the 18th century reported in his work 'Cancer of the larynx' two cases of what he considered to be cancers of the larynx, which by today's knowledge could have been other, nonmalignant lesions (Assimakopoulos 2003).

Significant changes in treatment approach have occurred together with the invention of the laryngoscope, in 1855 by the Spanish vocal pedagogist Manuel Garcia (Rosenberg 1971). The tool has been refined by others and implemented in clinical practice. Once the diagnostic component has been developed, the treatment has evolved and the first laryngectomies were undertaken with more or less success.

In the modern era, next to hypopharyngeal carcinoma, laryngeal cancer was one of the first candidates for fractionated external beam radiotherapy. Today, radiotherapy with concurrent cisplatin is the standard treatment for patients with locally advanced laryngeal cancer. Patients with complete response to induction chemotherapy were shown to have a high probability of cure after hyperfractionated radiotherapy (Majem 2006). While surgery is debilitating, advanced stage (T4) laryngeal cancers, particularly with cartilage invasion, are managed with total laryngectomy and adjuvant radiotherapy (Wick 2016).

Targeted therapy as well as gene therapy promises to improve the outcome of laryngeal cancer treatment in the near future. A significant association between the expression of p53 and poor patient outcome was found in patients with laryngeal carcinomas (Nylander 2000). Since in most laryngeal cancers the function of p53 gene is down regulated Wang et al. (1999) have explored the potential use of p53 in gene therapy of laryngeal cancer. The group has introduced a wild-type p53 into a laryngeal cancer cell line via a recombinant adenoviral vector and concluded that the adenovirus-mediated antitumour therapy is effective in inhibiting laryngeal cancer growth. After conducting a phase I clinical trial they have proven the effectiveness of recombinant adenovirus p53 injection (gendicine) in reducing laryngeal cancer progression (Han 2003). The 5-year relapse free survival was 100%. Further trials employing the adenoviral vector (Adp53) have shown that gendicine in combination with radiation show a synergistic effect in HNC (Zhang 2005).

Treatment of the head and neck has evolved greatly over the decades. Today, most early stage disease can be effectively managed with surgery or radiation as a sole agent. However, patients with advanced yet operable tumours are managed by surgery and postoperative radio- or chemoradiotherapy. Combined modality treatment is the current norm for patients with locally advanced, unresectable tumours. The role of chemotherapy in unresectable HNC was shown to be critical since radiotherapy alone leads to a failure-free survival of only 25% (Brizel 1998) (see also Section 2.4). Irrespective of the anatomical location, HNCs are generally characterised by aggressive repopulation and tumour hypoxia. To overcome accelerated repopulation during treatment, conventional radiotherapy is being increasingly replaced by altered fractionation regimens. Consequently, radiation treatment is given either in smaller multiple fractions a day (i.e. hyperfractionated radiotherapy), or as an accelerated regimen with shortened overall treatment time (i.e. accelerated radiotherapy) and sometimes with increased doses per fraction (i.e. hypofractionated radiotherapy). HNCs with high hypoxic content also benefit from altered fractionation, through tumour reoxygenation. Normal tissue protectors, radiosensitisers, bioreductive drugs and antiangiogenic agents also play an important role in HNC treatment, as additions to the already traditional radio- and chemotherapy.

2.4 CHEMORADIOTHERAPY IN HEAD AND NECK CANCER FROM THE EARLY DAYS TO THE PRESENT

As described in details in Chapter 8, cisplatin remains the most commonly used chemotherapeutic agent for the management of HNC. Cisplatin has been clinically used, in isolation, since the 1970s. Because the success of cisplatin as a single agent was limited, over the decades the drug was used in a variety of forms of combined chemoradiotherapy. In the

TABLE 2.2 Common Chemotherapeutic Agents Used in Combination with Radiotherapy in the Management of HNC

Drug Class	Representative Agents	Cell Cycle Effect/Properties
Alkylating agents/Platinum compounds	Cisplatin Carboplatin Oxaliplatin	DNA adduct, cell arrest in G2 phase, angiogenesis inhibitor, hypoxic cell sensitiser
Antimetabolites	5-fluorouracil Methotrexate	Interference with DNA synthesis
Antibiotics	Mitomycin-C Bleomycin Doxorubicin	Interference with cell growth Hypoxic cell cytotoxin DNA damage
Taxanes/ Plant alkaloids	Docetaxel Paclitaxel	Cell arrest in the radiosensitive phases of cell cycle Inhibition of microtubule function for cell replication

1980s and early 1990s the most common regimen consisted of high-dose weekly cisplatin administration. Nevertheless, in the mid to the late 1990s the benefits of fractionation have reached the field of chemotherapy, thus daily low-dose infusions were introduced and outcomes have continuously improved (Jeremic 1997 & 2008; Marcu 2003).

For higher efficiency and in order to overcome drug resistance, cisplatin is commonly administered in combination with other agents, where some of the most frequently used drug classes are the antimetabolites, antibiotics and taxanes. Table 2.2 presents the most common drugs used in combination with radiation and their main properties as radiosensitising agents.

However, multi-agent chemotherapy combined with radiation often increases treatment toxicity, which is not always balanced by better tumour control. The improved locoregional control and the increased percentage of complete response among patients treated with chemoradiotherapy in the late 1990s and the beginning of the 21st century are mainly due to the technological progress and the better knowledge of squamous cell radiobiology. Thus, the employment of altered fractionation radiotherapy delivered by advanced treatment techniques combined with cocktail chemotherapy resulted in better outcome as compared to traditional treatment regimens (Overgaard 2003; Bourhis 2006; Saunders 2010). Often the aggressive treatments resulted in more pronounced normal tissue toxicity, which however, could be managed. The higher tumour control is clearly indicative of effective treatment schedules, retaining cisplatin as the preferential chemotherapeutical agent for HNC.

2.5 HOPES AND FAILURES IN HEAD AND NECK CANCER RADIOBIOLOGY AND THERAPY

Tumour hypoxia, the epithelial growth factor receptor (EGFR), the VEGF and the mechanisms behind repopulation during treatment of HNC cells are only some examples of landmark discoveries that have shaped the treatment of HNC. There were also some high hopes raised by pre-clinical results of certain agents (such as tirapazamine, amifostine) that conveyed conflicting results when trialled (Figure 2.1). The sections below describe those agents and radiobiological processes that have shaped the treatment of HNC. While

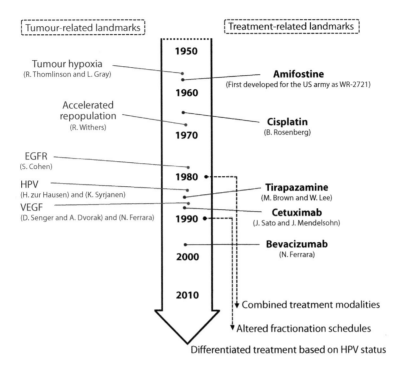

FIGURE 2.1 Hopes and failures in head and neck cancer radiobiology and therapy.

some of them were more successful in their clinical implementation than others, there were lessons learnt from each discovery, preclinical testing and clinical trial. There is no doubt that the science behind HNC oncology has progressed greatly and has shed light on several treatment-related aspects.

2.5.1 Tumour Repopulation and Altered Fractionation Schedules

Repopulation is one of the main factors responsible for treatment failure in HNC chemo-radiotherapy. For tumours undergoing rapid repopulation during treatment, such as squamous cell carcinomas of the head and neck, altered fractionation schedules have become widespread as they counteract the time factor. Altering conventionally frac-tionated radiotherapy to multiple fractions a day (hyperfractionated radiotherapy) or shortening the duration of treatment (accelerated radiotherapy) has been shown to limit tumour repopulation (Bourhis 2006). Hyperfractionated radiotherapy for advanced HNC within the RTOG 9003 trial resulted in a higher rate of locoregional control and no sig-nificant increase in late toxicity when compared with conventional treatment regimens (Beitler 2014).

Hypofractionation, a fractionation pattern which has been popular for other ana-tomical sites is now gaining room in head and neck radiotherapy, though sometimes for economical rather than radiobiological reasons. The theoretical advantage of larger-than-conventional doses consists in the counteraction of accelerated repopulation due to short-ened overall treatment time, which sometimes comes at the cost of late toxicities (Bakst 2011; van Beek 2016).

2.5.1.1 The Future of Fractionated Radiotherapy

The results of a meta-analysis reported by Bourhis et al. on 15 phase III trials, have established the advantage of hyperfractionation over both standard and accelerated radiotherapy (Bourhis 2006). Despite the superior locoregional control with altered fractionation, a large number of centres across the world adopt the conventional 2 Gy fractionation as standard of care. Shortening of treatment time via hypofractionation is an alternative to conventional treatment regimens unless this comes at the expense of unacceptable toxicities.

2.5.2 Therapeutic Ratio and Cisplatin

There are several ways to increase the therapeutic ratio in head and neck oncology and one of them is the multimodality treatment (Marcu 2013). Chemoradiotherapy is a standard approach in the management of locoregionally advanced HNC as the combined drug-ionising radiation has been proven over the last decades to lead to superior outcome when compared to radiotherapy as a sole agent. This is supported by a meta-analysis based on 93 randomised trials conducted between 1965 and 2000 on 17,346 patients, looking at the role of chemotherapy in HNC (MACH-NC) (Pignon 2009). Of all chemotherapeutic agents, cisplatin remains the most commonly administered drug, often in combination with others. However, given the adverse effects, normal tissue toxicity is a dose-limiting factor in combined chemoradiotherapy.

2.5.2.1 The Future of Cisplatin

Cisplatin is a highly potent drug and the oldest in the platinum family. Due to its toxic side-effects and also to overcome cisplatin-induced drug resistance (McWhinney 2009; Dasari 2014), there were several attempts to replace cisplatin with its sister platinum compounds: carboplatin, oxaliplatin, nedaplatin and mitaplatin. Despite the efforts to substitute cisplatin with other platinum compounds, this drug is still the most commonly administered chemotherapeutic agent in HNC and will probably stay as such for the years to come. Carboplatin might slowly gain more room, as based on the latest meta-analysis on definitive chemoradiotherapy for head and neck squamous carcinoma which evaluated cisplatin against carboplatin, the results have indicated no difference in response rate (Aguiar 2016).

2.5.3 Tumour Hypoxia and Tirapazamine

Tumour hypoxia is one of the main reasons for treatment failure, given that hypoxic cells are about three times more radioresistant to radiation than well-oxygenated cells. Advanced HNCs are commonly hypoxic, feature that is associated with poor prognosis (Nordsmark 2005). Out of several methods trialled for hypoxic cell sensitisation, the bioreductive drug tirapazamine was in the spotlight of head and neck oncology for a few decades due to its clinical success when combined with other anticancer agents. The interplay between tirapazamine and cisplatin was particularly remarkable through the repair inhibition or delay of cisplatin-caused DNA adducts (Brown 1999). Several trials on HNC have included tirapazamine as a hypoxic cytotoxin with varied results that were dictated by patient selection, HPV status and the complexity of hypoxia targeting (Le 2012). Nimorazole, an oxygen mimetic, antibacterial compound with the ability to sensitise hypoxic cells to the effect of radiation, has been trialled in HNC patients with more success (Overgaard 1991). Additional information on this drug can be found in Chapter 4.

2.5.3.1 The Future of Tirapazamine

The suboptimal results with tirapazamine in clinical studies were also due to the poor tumour penetration, given the fact that the drug is metabolised too fast in order to be optimally taken up by the hypoxic cell (Hicks 2004). Nevertheless, this limitation has been addressed in the second-generation of hypoxia-activated prodrugs. The tirapazamine analogue SN30000 has been designed with considerations of extravascular transport leading to more efficient tumour cell penetration and killing (Hicks 2010). Among the improved properties of SN30000 the following are to be mentioned: a threefold therapeutic gain as compared to tirapazamine on xenograft models and superior results when combined with fractionated radiotherapy (Hicks 2010). The preclinical test results of SN30000 warrant further research in HNC. Another prodrug, which has passed preclinical evaluation and currently is a phase III trial candidate in pancreatic cancer and soft tissue sarcoma is evo-fosfamide (TH-302) (Hunter 2015). While the novel hypoxia-activated drug is under investigation for several tumour sites, its value in HNC is yet to be demonstrated.

2.5.4 Epidermal Growth Factor Receptor and Cetuximab

Through promotion of epidermal cell growth and regulation of cell proliferation, the EGFR plays an essential role in HNC development, growth, metastatic spread and angiogenesis. Overexpression of EGFR was clinically proven to lead to increased tumour proliferation. About 90% of head and neck squamous cell carcinomas exhibit overexpression of EGFR (Kalyankrishna 2006), which could explain their high proliferative ability and aggressiveness. Furthermore, a recent study showed that EGFR regulates cell survival not only in the bulk tumour but also in the clusters of circulating tumour cells that can further lead to metastatic spread (Braunholz 2016). Elevated levels of EGFR correlate with poor prognosis, a fact that raised the need for the development of anti-EGFR agents. Cetuximab is a monoclonal antibody that was designed to interact with EGFR and inhibit the function of the receptor. The role of cetuximab in clinical settings is still controversial. A review of randomised phase III trials has indicated that while in combination with radiotherapy cetuximab has improved locoregional control in HNC patients, when added to platinum-based chemoradiotherapy no further improvement has been observed (Specenier 2013). Given that cisplatin-based chemoradiotherapy is the standard of care in HNC oncology, this result is not supportive of the routine implementation of cetuximab. Furthermore, recent results of a retrospective study in Asian population that compared cisplatin-based radiotherapy with cetuximab-based radiotherapy in locally advanced HNC, have indicated that cisplatin is associated with superior outcome (3-year overall survival 74% vs 42%) without increasing normal tissue toxicity (Rawat 2017). Nevertheless, recurrent and metastatic HNCs did benefit from the addition of cetuximab to chemoradiotherapy by prolonging patients' survival. This latter observation is confirmed by the study derived from the EXTREME trial results investigating the HPV status and its impact on outcome in HNC patients with recurrent/metastatic disease receiving chemotherapy with or without cetuximab (Vermorken 2014). The study has shown that the addition of cetuximab to platinum-based chemotherapy has improved survival in this patient group regardless of the HPV status, although patients that tested positive for HPV had longer survival than non-HPV patients.

2.5.4.1 The Future of Cetuximab

The usually lower incidence of cetuximab-associated normal tissue toxicity as compared to cisplatin-induced adverse events and the need to elucidate the role of cetuximab in HPV+ patients has led to the design of phase III trials that randomise HPV+ patients to receive either cisplatin or cetuximab with concurrent radiotherapy. Three such trials are currently ongoing: RTOG 1016 (US), De-ESCALaTE HPV (UK) and TROG 12.01 (AUS). Their primary goal is to assess and compare the toxicities in the two arms and to evaluate whether substitution of cisplatin with cetuximab will result in similar locoregional control and overall survival. Given the different treatment response of HPV– patients, a parallel trial would be needed for this patient group. Until then, the standard of care for unresectable HNSCC continues to be chemoradiotherapy (Specenier 2013).

Furthering the idea of patient stratification, a recent study suggests that adequate selection for cetuximab-based therapy can be critical as the pharmacokinetics (global clearance) of this monoclonal antibody influences overall survival in HNC patients (Pointreau 2016).

While other EGFR signal inhibitors trialled in the clinics, such as panitumumab (Mesia 2015) have been shown to be less efficient in triggering antitumour mechanisms, cetuximab remains in focus.

2.5.5 Vascular Endothelial Growth Factor and Bevacizumab

Another critical issue in the management of HNC is metastasis. It is a well-known fact that angiogenesis is a prerequisite for tumour expansion and promotion of metastatic spread. Proliferation of vascular endothelial cells and formation of new blood vessels are stimulated by angiogenic growth factors. The fact that tumour cells secrete a vascular permeability factor (today known as VEGF) that promotes angiogenesis was first observed in the 1980s (Senger 1983; Leung 1989). Once the role of VEGF was elucidated, this protein opened new research avenues to target VEGF. Bevacizumab (Avastin) is a humanised monoclonal antibody that targets VEGF and is clinically employed as an anticancer agent by its ability to block angiogenesis (Ferrara 2004). In head and neck carcinomas, bevacizumab is often combined with cisplatin-based chemo/radiotherapy. Both pre-clinical (Wang 2010) and clinical results (Fury 2012) show that the monoclonal antibody potentiates the effect of standard chemoradiotherapy.

2.5.5.1 The Future of Bevacizumab

Just recently, a phase III trial sponsored by the National Cancer Institute (NCI) has finished accruing patients with recurrent or metastatic HNC to be treated with platinum-based chemotherapy with or without bevacizumab with the primary objective to compare the overall survival between the two arms (clinical trials NCT00588770). Comorbidities, treatment-associated toxicities and progression-free survival are set as secondary objectives in this currently ongoing trial. The role of bevacizumab in HNC continues to be heavily investigated although with conflicting results. Fury et al. reported satisfactory results when bevacizumab was combined with cetuximab and cisplatin concurrently with IMRT in advanced stages of HNC. Thus the 92.8% 2-year overall survival and 88.5% 2-year progression-free survival, with tolerable toxicities are noteworthy results (Fury 2016). A phase I trial of bevacizumab with concurrent cisplatin-based chemotherapy confirmed the potential of bevacizumab in reducing tumour proliferation and hypoxia (as assessed by FLT-PET and Cu-ATSM PET), thus increasing the efficacy of chemotherapy in HNC patients (Nyflot 2015). On the other hand, in a phase II randomised trial of radiotherapy combined with cetuximab, pemetrexed with or without bevacizumab in locally advanced HNC, the addition of the VEGF targeting agent resulted in increased toxicity without any benefit to the tumour (Argiris 2016). All these recent studies demonstrate the increased interest in elucidating the role of this monoclonal antibody in the management of HNC, reiterating the importance of patient stratification and treatment individualisation.

2.5.6 Normal Tissue Protection and Amifostine/Radioprotectors

Despite significant advances in treatment, long-term survival among HNC patients has not improved. One major drawback of the aggressive therapy is normal tissue toxicity. Mucositis, xerostomia, myelopathy and pneumonitis are some of the most common dose-limiting adverse events in head and neck radiotherapy. Acute side-effects can usually be managed with appropriate medical care; however, late toxicities are often irreversible thus in need for better organ sparing or dose de-escalation. While the latter is not possible without hindering tumour response (at least for HPV− HNC), sparing of healthy tissue can be achieved with careful treatment design and/or the addition of an agent that could offer normal tissue protection and increase quality of life. One such agent that has been trialled to partially reduce side-effects in healthy organs during HNC radio/chemotherapy is amifostine, a prodrug that has preferential uptake by normal cells. However, the results of clinical trials employing amifostine were inconclusive and the reduction in normal tissue toxicity was not always demonstrated (Marcu 2009). Some of the best protected organs against the effect of radiation were shown to be the salivary glands and the hematopoietic system, both by a protection factor of three in terms of toxicity (Kouvaris 2007). Nevertheless, since amifostine has no ability to cross the blood-brain barrier, the central nervous system gained no benefit from this agent. To date, there is still no unanimous evidence for the efficacy and safety of amifostine. However, a recent meta-analysis based on randomised controlled trials for HNC that included amifostine could statistically confirm the clinical benefit of this radioprotector (Gu 2014).

2.5.6.1 The Future of Radioprotectors

To overcome the limitations of amifostine, several radiation protection agents and radiation mitigators are under evaluation. Some of the promising agents with radioprotective effect are nitroxides, with *tempol* being the leading agent that fulfils the requirements of a radioprotector (Citrin 2010): (1) to scavenge free radicals, (2) to exhibit antioxidant properties and (3) to show good tissue selectivity. Other possible candidates to play the role of radioprotectors that have been tested in randomised trials are the antioxidant vitamins (Bairati 2005). However, they did not pass the tests, as their presence was associated with poorer tumour control.

Radiation mitigators are also in focus. These are the agents that could either prevent acute toxicity by interfering with the post-irradiation processes or could repair the injured tissues by stimulating repopulation in the affected cell compartments (mucosa, bone marrow) (Citrin 2010). One such agent is palifermin that was shown to reduce severe mucositis in HNC patients treated with cisplatin-radiotherapy (Le 2011). To establish the definitive role of radiation mitigators in the management of HNC, more research is warranted.

2.5.7 HPV-Associated Head and Neck Cancer

The discovery of HPV-related HNC in 1983 (Syrjänen 1983) is considered a milestone in head and neck oncology as it has introduced a need for a new approach in HNC management. Several trials have reported superior results for HPV+ cancers as compared to HPV− HNC. Results from the Danish Head and Neck Cancer group (DAHANCA) show that while the presence of HPV and p16 overexpression represent a positive and independent prognostic marker of outcome for oropharyngeal cancers, positive testing for HPV in other HNC (such as laryngeal or hypopharyngeal carcinomas) does not warrant the same high response to treatment (Lassen 2014).

2.5.7.1 The Future of HPV-Associated HNC Treatment

Currently the treatment for head and neck carcinomas is not differentiated as a function of HPV status. However, a new treatment approach is required because of the higher radioresponsiveness of HPV+ HNC, which could redesign the current treatment regimens through dose de-escalation for better normal tissue sparing. The observation from the DAHANCA study, whereby HPV status-dependent response was found in oropharyngeal cancers only, will certainly influence future treatment guidelines for oropharyngeal cancers (Lassen 2014).

Personalised treatment planning and delivery is becoming the leitmotif of today's oncology. As the father of medicine wisely said: 'It is more important to know what sort of person has a disease than to know what sort of disease a person has' (Hippocrates).

CHAPTER 3

General Radiobiology Refresher

3.1 RADIATION QUANTITIES AND UNITS

The need to compare and reproduce clinical outcomes and also to protect against the unwanted effects of ionising radiation has led to the introduction of a number of concepts and quantities to characterise and quantify the amount of ionising radiation. A central quantity in this respect is the dose of radiation. The terminology originates in the medical applications of ionising radiation and is an analogy to the doses of medicines from pharmacology leading to certain effects. Nevertheless, the dose of radiation is a physical quantity and the physical methods developed to describe it have been proven in time superior to the biological methods to characterise the effects of radiation.

The absorbed dose of radiation represents the mean energy imparted by ionising radiation to matter per unit mass. In the International System of Units (SI), the unit of measure for dose is Gray (Gy) representing Joules (J) per kilogram (kg). In the centimetre-gram-second (CGS) system of units, the unit of measure for dose is the rad, which represents 0.01 Gy. This is a unit that has been much used historically and some clinics still report doses in cGy to maintain a link with their rad-experience. Nevertheless, the use of the SI unit, Gy, is used almost universally in Europe, North America and elsewhere.

Energy deposition in tissues takes place through the interaction processes suffered by radiation. It is beyond the purpose of this book to go into the details of radiation interaction processes. Nevertheless, the type of interactions gives one possible classification of ionising radiation. Thus, charged particles interact directly with the atoms in the tissues and transfer continuous energy to the matter, leading either to direct local deposition of energy or the creation of secondary particles that are subjected to further interactions, this behaviour being characteristic of directly ionising radiation. In contrast, photons, neutrons and other uncharged particles interact with matter through a few catastrophic processes in which secondary charged particles are created that interact further with the atoms in the tissues, this behaviour being characteristic of indirectly ionising radiation. The average energy imparted to the medium per unit mass in the interaction processes suffered by the particles represents the absorbed dose. Other quantities, like kerma or collision kerma, characterise the energy transferred or the net energy transferred in the interaction processes in the tissues.

The energies involved in dose deposition are rather small in absolute values. Thus, a dose between 4 and 5 Gy delivered uniformly to the whole body is considered a median lethal dose for humans. Assuming an average human of 70 kg (Walpole et al. 2012), the mean lethal dose of 5 Gy results from an energy deposition of 350 J that could barely raise the temperature of the whole body. Furthermore, radiation treatments deliver doses sometimes in excess of 70 Gy to limited volumes of tumours with limited damage to the surrounding normal tissues. For a tumour target with a volume of 100 cm^3, this dose results from an energy deposition of only 7 J (assuming the density of water for the tissue), but this is enough to cure many tumours. These comparisons indicate that not only is the energy deposited important for biological effects but also where and how this energy is deposited; more details about this will be given in the following sections.

A historical quantity of interest is radiation exposure and it represents the mean ionisation created by photons per unit mass of air. As such it is less useful than the absorbed dose to characterise radiation effects in tissues; however, it is sometimes used to characterise the strength of radioactive sources. The SI unit for exposure is C/kg, although the roentgen (R) has also been used in the past. Conversion tables are available to link radiation exposure to doses in materials of interest, but these are generally valid for conditions where radiation equilibrium is met and therefore not useful for heterogeneous irradiations.

Radioactive sources are also characterised in terms of activity, representing the number of decays or nuclear transformations per second. Activity is much used in nuclear medicine, for example, to characterise the amount of radioactive material given to the patient. The SI unit for activity is the becquerel (Bq) and is defined as one nuclear transformation or decay per second. An older unit for activity has been the curie (Ci), which has been defined based on the activity of 1 g of radium, as 3.7×10^{10} Bq, although the use of this and other non-SI units is strongly discouraged. Tables also exist for the conversion of activity into the exposure rate, taking into account the source type and the irradiation geometry. However, the extrapolation of the conversion is not easy when the irradiation or the medium becomes heterogeneous.

The study of energy deposition in matter by directly or indirectly ionising radiation, as well as the relationships between various quantities and irradiation geometries, make the topic of radiation dosimetry. This is a physical science that is also concerned with the development of methods for the quantitative determination of energy depositions in media by ionising radiation. However, it should be recognised that although there is no simple relationship between biological effect and dose, the latter is used as a substitute for the former due to the reliability of the methods used to quantify and characterise it.

3.2 RADIATION ACTION AND RADIATION DAMAGE

Early experiments performed with short range alpha particles (Munro 1970) have shown that the targets for cell killing were either nuclear or in the immediately perinuclear cytoplasm, leading later to the conclusion that the critical structure for radiation effects in biological systems is the deoxyribonucleic acid (DNA) molecule. This molecule has a complex structure and contains the genetic information in the cells. The backbone of the molecule is a polymer of phosphate and sugar bases on which the genetic information is coded in a unique sequence of four nucleobases: adenine (A), cytosine (C), guanine (G) and thymine (T), which are bound to the sugar-phosphate chain. In living cells, the DNA does not exist as a single molecule, but as a double helix of pairs of molecules bound through hydrogen bonds between nucleobases that pair in a unique way, adenine with thymine and guanine with cytosine. The pairing not only allows the stabilisation of the molecule in the double helix with a diameter of about 2 nm but also the maintenance of two copies of the genetic codes that are used for DNA replication and repair. Depending on the phase of the cell cycle or the repair status, the DNA double helix can be found either as a loose chromatin structure or in more condensed chromosome structures with higher order coiling, which is also associated with increased mechanical stress on the molecule.

Radiation could interact directly, either with the atoms of the DNA molecule or with those of the surrounding molecules, mainly water. The latter interactions lead to the formation of chemically unstable free radicals that could further interact with the double strand of the DNA. These two interaction pathways, direct and indirect, represent the two modes of action of radiation. It should be noted that these modes of action have no relationship to the types of ionising radiation; the identical nomenclature is just a coincidence. In this context, it is important to mention that the proportion of direct and indirect modes of action is a characteristic of radiation type. Thus, for photons and electrons, the indirect mode of action is the dominant one, which is responsible for about two-thirds of the damage to the DNA. In contrast, the direct mode of action becomes dominant for particles with a higher ionisation potential.

The most important interactions from the perspective of subsequent biological events are those that involve a sufficient energy transfer to break chemical bonds in the molecules. It is interesting to note that only a small fraction of the total interactions results in such transformations, the majority of interactions leading to excitations and vibrations that are ultimately transformed into heat with little impact upon the status of the chemical bonds.

The initial damage to the DNA molecule that leads to biological effects is a break in one of the two strands of the double helix molecule. Typically, this is called a single strand break (SSB). If two SSBs are in close proximity on the opposite strands of the double helix, the damage is called a double strand break (DSB). Further clustering of strand break leads to the formation of multiple damaged sites with increased complexity of damage to the DNA molecule. The clustering of damage could be the result of SSB clustering and also of the coiling of the DNA molecule and the strain that exists in the molecule. This indicates that the production and severity of damage depends on both the density of ionisation events and the cell cycle phase.

It is important to recognise that not all the initial biochemical damage leads to cellular and tissue effects due to the existence of repair mechanisms. Indeed, the induction of damage to DNA by radiation or other agents triggers repair mechanisms that attempt to restore the integrity of the molecule. The repair mechanisms vary depending on the type of damage. Thus, SSBs are typically repaired through excision repair mechanisms that remove the damaged nucleotide and replace it using the undamaged DNA strand of the double helix as a template for matching the suitable nucleotide. In the case of DSBs, repair mechanisms either attempt to join the broken ends of the DNA double helix as is the case of non-homologous end-joining (NHEJ) or use a sister template of the DNA double helix to restore the right sequence of nucleobases as in the case of homologous recombination (HR). Nevertheless, the repair processes are not equally effective and the differences in the repair mechanisms are also reflected in the fidelity of the repair process, with NHEJ representing a quick but low-fidelity repair mechanism, while HR is a slower but higher fidelity mechanism. Thus, if several DSBs have been produced, it is quite possible that the wrong ends of the DNA molecule are joined and following this process genetic information is lost or modified.

The complexity of damage to the DNA is in inverse relationship to the capacity of the cell to repair it and therefore more complex clusters of damage are less likely to be repaired. Furthermore, there are a number of genetic mutations that lead to impaired or dysfunctional repair mechanisms and they are associated with increased sensitivity to radiation.

If the damage to the DNA is correctly repaired, there are no future consequences for the cell. However, unrepaired or incorrectly repaired damage has important implications for the cells as it leads to genetic mutations as the result of changes or losses of genetic material. From this point of view, further distinction could be made between viable and lethal mutations, the former being responsible for cellular dysfunctions and the latter leading to cellular death. Thus, viable mutations in the somatic cells are associated with stochastic effects including cancer induction, while mutations in the germ cells are passed to the progeny and could cause deleterious effects in the offspring. In contrast, cell death is associated with deterministic effects in tissues. Cell death is a generic term including the loss of the reproductive capacity by the cell, but also programmed cell death through apoptosis or senescence. Which cell death path is chosen, depends on the type of mutation that has been produced. Thus, incorrect joining of the DNA could lead to aberrations that could prevent the cell from undergoing the mitosis process. Other mutations could be recognised by the systems checking the integrity of the DNA and trigger the sequence of events that lead to the programmed death of the cell.

These effects are also the basis of effects in higher order systems. Organs and tissues are highly organised structures based on functional subunits (FSU) and reserve capacity that perform a specific function. Nevertheless, cell death in these structures would ultimately lead to impaired organ or tissue function and some of these effects could even be life-threatening.

3.3 DOSE RATE EFFECT, LET AND RBE

Establishing the relationship between dose and biological effect has to take into account both the induction of lesions and their repair. Thus, for a given type of radiation, the creation of lesions like SSBs and DSBs depends linearly with dose (Rothkamm & Löbrich 2003). However, the biological effects observed in cells and higher order systems are the result of remaining lesions after the repair process and therefore their dose dependence is modulated by the accumulation or clustering of the lesions. Clustering could appear at high doses of radiation when several lesions are produced more or less simultaneously, when lesions accumulate at a higher rate than could be removed by repair or when there is a high density of ionisation events in or near the DNA molecule. Lesion accumulation and interaction at high doses has been used in theories of radiation action (Kellerer & Rossi 1972; Chadwick & Leenhouts 1973) to explain the dose response relationship for biological effects. Similarly, the temporal imbalance between lesion induction and repair also modulates the appearance of biological effects. Thus, at high-rate irradiations it is possible that before a lesion is removed by repair, a second lesion could be produced in its proximity leading to a complex lesion that is more difficult to repair. This modulation of the dose response due to the temporal interplay between lesion induction and repair has been dubbed the dose rate effect. The dose rate effect has been studied in cells in vitro and it predicts that the same dose of radiation leads to less biological effect if it is delivered at a lower rate, as repair could remove some of the lesions before they are set by interactions with additional lesions. The derived relationships for the dose fractionation and dose rate effects have been subsequently validated in systems in vivo and they are nowadays very much used for clinical treatments to optimise the balance between tumour and normal tissue effects.

The ionisation potential of radiation is another factor that modulates the biological effects. This potential is quantified as a linear energy transfer (LET) that gives a measure of the energy transferred by the particles per unit length of their track. For directly ionising particles, the LET is the stopping power restricted only to transfers that could be considered local, in the proximity of the DNA molecule. For indirectly ionising radiation, it represents an average of the restricted stopping power of the secondary particles created from interactions with the medium. However, due to the stochastic character of the interaction of these particles, one could distinguish between track average LET and energy average LET, depending on how the average has been calculated. This distinction however is most prominent for particles with high LET that have a large variance of the energy deposition events and to a lesser extent for low LET radiation. Photons and electrons that are the most used radiation modalities in modern radiotherapy are regarded as low LET radiation, while neutrons and some of the ions used for radiotherapy are regarded as high LET radiation. It should be noted however that there is no threshold for the transition from low to high LET and that even for the same type of radiation the LET varies with its energy and even the medium, in the same manner as the variation of the stopping power of the charged particles involved.

Low LET radiation is generally characterised by sparsely ionising tracks with a low potential for producing DNA lesions of high complexity. These in turn are most likely produced from the clustering of lesions as the result of accumulation of particle tracks in the same region of the DNA molecule or from interactions with free radicals produced through water radiolysis in the proximity of the DNA molecule. In contrast, high LET radiation is characterised by densely ionisation tracks that have an increased potential to produce complex lesions. Thus, for high LET radiation, fewer tracks are required to produce the same number of lesions as low LET radiation. Furthermore, this could be achieved by the direct interactions with the DNA molecule, with less contribution of the free radicals, thus changing the balance between the direct and indirect modes of action for producing lesions to the DNA. In other terms, for the same dose of radiation (energy deposited in a given volume), more lesions with increased complexity are produced by high LET radiation than by low LET radiation, which in turn lead to more biological effects of the former compared to the latter. This increased effectiveness of high LET radiation is quantified as the relative biological effectiveness (RBE) which is defined as the ratio of doses of two radiation types which lead to the same biological effect. Due to the widespread use of photons in radiation therapy, they are usually the reference radiation for RBE comparisons, in particular the 250 kVp X-rays or Co-60 gamma radiation. In these conditions, the RBE of a given type of radiation is the ratio of photon dose leading to a certain effect to the dose of the tested radiation leading to the same biological effect, all other factors being the same. Given the complex relationship between radiation damage and biological effect, the RBE varies both with the endpoint and the tissue type and these variations have to be accounted for when extrapolating dose relationships from photons to other types of radiation. Nevertheless, the general trend is that the RBE increases with the LET of the radiation, since an increased ionisation density increases the probability of creating DSB of the DNA molecule. In this context, a maximum effect for the RBE is encountered for radiation that has the mean distance between ionisation events comparable with the diameter of the DNA molecule. However, for even higher LET, the RBE starts to decrease due to an overkill effect where energy depositions can take place in between the strands of the double helix and thus cannot contribute to the creation of further damage. The increased potential of high LET radiation to create complex DNA lesions gradually decrease the importance of the indirect mode of action, and, therefore, the direct mode of action becomes the dominant mode that eventually leads to biological effects.

3.4 THE OXYGEN EFFECT

The oxygen effect is the name given to the modulation of the biological effects of radiation by the presence of oxygen. Although several factors are known to modulate the response of tissues to radiation, some of these being presented in the previous sections, the oxygen effect has a prominent role in radiation biology due to the physiological importance of oxygen for living tissues and the magnitude of the modulation which is one of the largest in radiation biology. In this context, it is important to note that the oxygen effect does not originate in any metabolic or physiological effect of the oxygen, but simply in the electron affinity of the oxygen molecule that makes it one of the most reactive chemical species. This chemical affinity enables oxygen to act on the free radicals formed by radiation and through them to potentiate the indirect effect of radiation (Hall & Giaccia 2011) or to fix-ate existing damage to the DNA molecule (Prise, Gillies & Michael 1999). Since the nature of the chemical reactions involved in these processes, the presence of oxygen is required during or at most a few milliseconds after irradiation, during the lifetime of the free radi-cals created through the indirect mode of action of radiation. This has been demonstrated in advanced experiments allowing oxygen to rapidly mix with anoxic cells in timing-controlled microexplosions at various times before or after very short pulses of radiation (Michael et al. 1973).

The oxygen effect has been observed in parallel with the use of radiation on biological systems. Thus, in the beginning of the twentieth century, Hahn and Swartz reported that the effect of radium applicators on the skin decreases when pressure is applied (Hahn 1904; Schwarz 1909). The effect has subsequently been univocally related to the presence of oxy-gen during irradiation (Crabtree & Cramer 1933). Systematic experiments in the following years have established that the dependence of the effect on oxygen concentration is non-linear (Alper & Howard-Flanders 1956). Thus, the full magnitude of the effect is seen in essentially anoxic conditions and fairly low amounts of oxygen, around 20 mmHg partial pressure, are needed to achieve full radiosensitisation that remains essentially the same for rather broad ranges of oxygen concentrations.

For irradiations with photons and other low LET radiation, doses that are 2.5–3 times higher are needed in the absence of oxygen (anoxic conditions) to achieve the same effect as in the presence of oxygen (oxic conditions). The ratio of doses administered in anoxic and respectively oxic conditions to cause the same effect has been termed the oxygen enhance-ment ratio (OER) and represents a measure of the magnitude of the oxygen effect. The OER for photon irradiation has been confirmed in a broad range of experiments employing var-ied biological systems and endpoints. It should be mentioned that the relationship between the oxygen effect and the indirect mode of action is also supported by the decrease of the OER for higher LET radiation (Figure 3.1), for which the direct mode of action becomes increasingly important. Indeed, experiments performed with neutrons and heavier ions have shown that the oxygen effect could be significantly decreased or even abolished as the LET of the radiation increases.

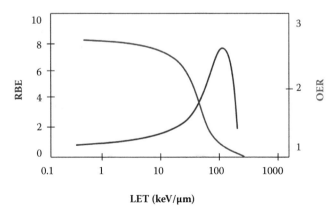

FIGURE 3.1   RBE and OER as a function of LET.

The oxygen effect has profound implications for radiation therapy. Thus, starting with the 1930s, tumours have been shown to contain regions with impaired vasculature (Mottram 1936; Thomlinson & Gray 1955) that leads to the development of hypoxia, which, in turn, affect the radiosensitivity of the tumour cells (Gray et al. 1953). Later studies have shown that tumour vasculature is inadequate to meet the demands of the developing neoplasm and as it originates on the venous side of the vasculature also has blood with low-oxygen concentrations (Vaupel, Kallinowski & Okunieff 1989; Denekamp, Dasu & Waites 1998). All these aspects made hypoxia; a common feature of tumours that increases their resistance. This is in contrast with normal tissues where a suitable vascular network exists to provide an adequate supply of high-quality blood which in turn makes all of the cells well oxygenated and consequently sensitive to radiation. This difference in radiosensitivity between tumour and normal cells represents a practical conundrum in radiation therapy and several approaches have been developed to overcome it as illustrated in Chapter 4.

3.5 PROLIFERATION KINETICS AND PARAMETERS

Cancer is considered a genetic disease involving abnormal cells that have escaped growth control and have gained a series of additional capabilities that allow them to become independent and invasive in comparison to normal cells (Hanahan & Weinberg 2000; 2011). From this perspective, it is important to characterise the proliferation features of tumours.

Both normal and tumour cells replicate via mitotic division, the process through which the cells give rise to two identical daughter cells. Technically speaking, mitosis (M) is a phase in the development of the cell in which replicated DNA condenses into chromosomes which in turn separate to form two identical copies for the progeny cells. The time between two successive mitoses represents the cell cycle characterised by a series of well-defined events. One of these is the synthesis (S) of copies of the DNA molecules in preparation of the mitotic phase. Studies of the cell cycle have shown that between the M and S phases are two periods of apparent inactivity, termed gaps. The first gap, G_1, describes the interval between M and S, while the second gap, G_2, describes the interval between S and M. It is important to note that the amount of DNA in the cell varies along the cell cycle. Thus, in G_1 the cell contains only one copy of the DNA, while in G_2 the cell nucleus contains two identical copies of the DNA. The varying amount of DNA and the condensed versus loose state in which could be found across the cell cycle determines some variations in radiosensitivity between cell cycle phase. Thus, the S phase is the most resistant as the DNA is in the loosest state and is surrounded by molecular complexes that check its integrity during duplication, while late G_2/M is the most sensitive since the DNA is highly condensed and subject to mechanical stress that could easily result in complex lesions.

The progression between the phases of the cell cycle could be studied in so-called labelling studies in which a precursor to the synthesis of the DNA, e.g. thymidine or another nucleobase analogue, which has been labelled with a suitable tracer is made available to the cells. If the exposure is brief, only the cells in the S phase will incorporate it in their DNA and carry it through their cell cycle. Initial studies have been made with radioactive tracers, but subsequent ones have used fluorescent compounds for increasingly complicated flow cytometry analyses of the proliferation kinetics of the cells that look at the fraction of labelled cells in relation to unlabeled cells or their intrinsic content of DNA. Thus, the progression of the cells through successive mitoses describes the cell cycle length (T_C). Similarly, the time needed for the labelled cohort to reach mitosis is the length of the G_2 phase, while the time needed for all labelled cells to go through mitosis gives the length of the S phase and so on. The characteristic times for the various phases are related through algebraic operations and one could determine the length of various phases if the other ones are known, since $T_C = T_{G1} + T_S + T_{G2} + T_M$. However, experimental determinations of the cell cycle parameters are more complex due to the need to account for the heterogeneous distribution of the cells around the cell cycle and the fact that cells do not progress synchronously through various phases.

Nevertheless, the technique has been proven reliable to define a series of parameters that could be used to characterise cell proliferation. Thus, the fraction of cells that are in the M phase relative to all proliferating cells, the mitotic index (MI), has been related to the ratio between the length of the mitosis, T_M and the cell cycle length, T_C (Equation 3.1).

$$MI = \lambda \cdot T_M / T_C \qquad (3.1)$$

where λ is a parameter accounting for the distribution around the cell cycle, e.g. assuming an exponential distribution, $\lambda = \ln(2)$. Similarly, a labelling index (LI) could be defined as the fraction of cells in the S phase relative to all proliferating cells and could be related to the corresponding characteristic times of the cell cycle (Equation 3.2).

$$LI = \lambda \cdot T_S / T_C \qquad (3.2)$$

Variations in the lengths of cell cycles could be explained from the variability in the length of the G_1 phase, while the lengths of the other phases are surprisingly similar. Moreover, it has been shown that in many populations, only a fraction of cells go actively through the cell cycle, while the remaining cells are in a dormant state analogous to G_1, termed G_0, either due to inhibiting signals or due to lack of nutrients needed for the synthesis and division processes. Nevertheless, these cells could be recruited back in the cell cycle.

Based on these observations, one can define the growth fraction as the fraction of cells actively going through the cell cycle relative to the total number of cells (Equation 3.3).

$$GF = P/(P + NP) \qquad (3.3)$$

where P is the number of proliferating cells and NP is the number of non-proliferating cells.

The growth fraction could also be used for defining the potential doubling time of the population, T_{pot}, which represents the time needed for the whole population to double in size in the absence of cell loss (Equation 3.4).

$$T_{pot} = T_C / GF \qquad (3.4)$$

T_{pot} could also be defined based on an LI much used in flow cytometry, LI_{FCM}, relating the number of labelled cells to the total number of cells, proliferating and non-proliferating (Equation 3.5).

$$LI_{FCM} = \lambda \cdot T_S / T_{pot} \qquad (3.5)$$

Clinical growth of tumours takes place at a much slower rate than predicted from either T_C or T_{pot}. The explanation for this behaviour is that the overall growth of the tumours is not only the result of the proliferation of all or a fraction of cells, but also the balance between proliferation and cell loss. Indeed, cells could be lost from a tumour in a number of ways, including death through starvation due to inadequate vascular network, programmed death or metastasis formation. Consequently, a cell loss factor, Φ, has been defined (Steel 1968) to account for the difference between T_{pot} and the actual tumour doubling time T_D (Equation 3.6).

$$\Phi = 1 - T_{pot} / T_D \qquad (3.6)$$

A number of studies have attempted to use proliferation kinetic factors like T_{pot} as predictive markers for tumour progress and treatment outcome with various but limited degrees of success (Begg et al. 1999).

3.6 DOSE-RESPONSE RELATIONSHIPS

The main aim of radiotherapy, expressed in clinical terms, is the eradication of the tumour while sparring as much as possible of the normal, healthy tissue. In physical terms this aim is expressed as delivering the prescribed dose to the target while keeping the doses to the organs at risk and the normal tissue below the tolerance levels. In radiobiological terms this translates into maximising the probability of controlling the tumour while minimising the probability of normal tissue injury.

An important point in clinical radiobiology is therefore the characterisation of the relationships between radiation-induced cell death and the probability of eradicating the tumour cells as well as the relationship between the cell killing and the probability of injuring the normal tissue or losing the functionality of the healthy organs at risk. The focus is therefore on the relationship between the delivered dose of radiation and the consequent tumour and tissue response, as well as the factors affecting this response. The relationship between the absorbed dose and the probability of controlling the tumour or the probability of normal tissue complication is usually represented by a pair of sigmoid dose-response curves as illustrated in Figure 3.2. Thus, in order to achieve a high probability of tumour control without injuring the normal tissue, the two curves should be pushed away enlarging therefore the so-called therapeutic window. In modern radiation therapy, this is achieved by conforming the high-radiation dose to the target considering at the same time the radiobiological factors affecting the tumour and the normal tissue response.

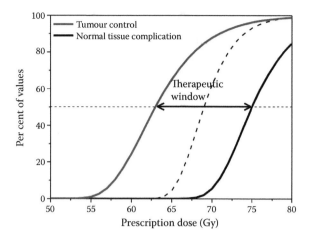

FIGURE 3.2   Illustration of the dose-response relationships for the probability of controlling the tumour and the normal tissue complication as well as the concept of therapeutic window.

The response of a tumour or an organ at risk to radiation can be estimated based on the number of rescued FSU. One could therefore assume a number of Tissue Rescue Units (TRU) as the minimum number of FSU that ensure the functionality of the tissue or organ. Thus, after irradiating the tumour or the normal tissue with a given dose resulting in a given surviving fraction (SF) of cells in that particular tissue, the number of surviving TRU would be the TRU·SF. The probability of tissue functioning failure would therefore be (1-SF)·TRU, which, for small SF and large TRU, could be approximated as exp(-TRU·SF). Assuming Poisson statistics, the probability of tissue failure is described as P = exp(-TRU·SF).

For tumours, the FSU are the clonogenic cells, meaning that all clonogenic cells in the tumour must be inactivated in order to achieve local control. For normal tissues, however, the response is governed by several factors such as the inherent cellular radiosensitivity, the kinetics of the tissue and the way the tissue is organised. Thus, normal tissues can be classified as serial, parallel, or mixed with respect to the way the FSU constituting them are organised while tumours could be regarded as the ultimate parallel structures.

The response of normal tissues can also be divided into early and late effects depending on the time of onset.

For further information on radiation biology aspects from subcellular to cellular and tissue level, readers are directed to relevant references such as Joiner & van der Kogel (2009), Hall & Giaccia (2012) and Sureka & Armpilia (2017).

CHAPTER 4

Hypoxia and Angiogenesis

4.1 TUMOUR HYPOXIA

Radiobiological hypoxia, defined as the lack of oxygen in the cellular tumour microenvironment, has been recognised for a long time as one of the critical factors that determine the cellular response to radiation (Thomlinson 1955). The role of cellular oxygenation as modulator of the radiation response has been observed and extensively studied *in vitro* (Gray 1953; Hill et al. 2015) as well as in solid tumours *in vivo* (Horsman 2017). For head and neck carcinomas in particular, tumour hypoxia has been proven to be a negative prognostic factor with respect to both local control and overall survival (Overgaard 2011) and; therefore, it is now regarded as one of the main problems in curing head and neck cancer (HNC).

The particularities of the tumour microenvironment and the presence of tumour hypoxia might be responsible for the failure of radiotherapy, chemotherapy and, to some extent, surgery, in solid tumours in general and in HNC in particular (Overgaard 1994). The poor tumour vasculature and the lack of a lymphatic system in tumours making difficult the drainage of the metabolic residues might lead to difficulties to efficiently deliver chemotherapeutic agents to all the cells in the tumour. Furthermore, the action of some of the chemotherapeutical cytotoxic agents might be hindered by the increased pH in the tumours. For radiation therapy in particular, however, the lack of adequate oxygen supply is by far one of the most important factors with respect to tumour response to radiation.

The generic term of tumour hypoxia refers to the impaired oxygen to the cells due to the deficient tumour vasculature in conjunction with the increased interstitial pressure leading to low values of the partial pressure of oxygen in the tumour microenvironment. The term hypoxia, however, is further refined in order to discriminate between the two main types of hypoxia depending on the way they originate in the tumour. Thus, depending on the underlying mechanism leading to the lack of oxygen supply to the tumour cells, one refers to diffusion or perfusion limited hypoxia.

For many years, it was thought that the only type of radiobiological hypoxia observed in tumours was solely the result of the limited diffusion of oxygen into tissue. Thus, when the distance between blood vessels in tumours becomes comparable to the distance of oxygen diffusion into tissue due to cellular consumption, the tumour cells become chronically deprived of oxygen leading to the occurrence of diffusion limited hypoxia, frequently referred to as chronic hypoxia. One of the first demonstrations of chronic hypoxia was made by Thomlinson and Gray in their combined study published in 1955 in which they showed that the width of the viable rims of cells measured from the periphery of the tumour well supplied with oxygen from the vascular stroma to the necrotic core had similar values with the calculated maximum distance of oxygen diffusion into tissues (Thomlinson & Gray 1955).

The presence of perfusion-limited hypoxia, also known as acute hypoxia, was not acknowledged until 1979 when Martin Brown suggested that local changes in blood flow determined by perfusion-related events such as the temporary occlusion of capillaries by red blood cells or the collapse of blood vessels because of high interstitial pressure might lead to transient regions of hypoxia (Brown 1979). The initial hypothesis by Brown was later confirmed and demonstrated in dual hypoxic markers experiments (Chaplin, Olive & Durand 1987; Bussink et al. 1999; Rijken et al. 2000).

For many years, it was thought that the only difference between chronic and acute hypoxia is the time scale, chronic hypoxia being, to a large extent, a long-lasting state, while acute hypoxia being transient lasting from minutes to hours; however, the two types of hypoxic cells have the same induced radioresistance. Recently, however, it was suggested that their sensitivity to radiation might be different. In particular, experimental evidence keeps gathering indications that chronically hypoxic cells are more sensitive than acutely hypoxic cells as the metabolic energy status of the cell might act as a modulator of the intrinsic radiosensitivity (Denekamp & Dasu 1999). It has also been postulated that, in some cases, chronically hypoxic cells might even be more sensitive than the oxic cells (Dasu 2000). In addition, there are several other studies showing that the genes involved in the molecular mechanism of DNA repair through homologous recombination are downregulated in cells exposed to chronic hypoxia compared to acute hypoxia leading to a lower oxygen enhancement ratio and hence to higher radiosensitivity (Sprong et al. 2006; Chan et al. 2008; Luoto, Kumareswaran & Bristow 2013; Chan et al. 2014). Defects in homologous recombination and mismatch repair have also been observed in several experimental studies in diffusion limited hypoxia (Meng 2005) while functional increase in non-homologous end joining is associated to perfusion limited hypoxia (Um et al. 2004). Furthermore, the loss of repair capacity of hypoxic cells with low ATP levels has been reported in several experimental studies starting from the early work of Hall and colleagues (Hall, Bedford & Oliver 1966) to the experiments by Zölzer and Streffer on several cell lines including human head and neck squamous carcinoma (Hall 1972; Nagle, Moss & Roberts 1980; Gerweck, Seneviratne & Gerweck 1993; Gerweck, Koutcher & Zaidi 1995; Pettersen & Wang1996; Zölzer & Streffer 2002). The occurrence of the two types of hypoxia in HNC and the main differences between them are illustrated in Figure 4.1.

Hypoxic cells can also acquire a mutator phenotype leading to decreased DNA repair, increased chromosomal instability as well as an increased mutation rate (Bindra et al. 2004; Bindra & Glazer 2005). Thus, the differences in DNA repair capacity of chronically

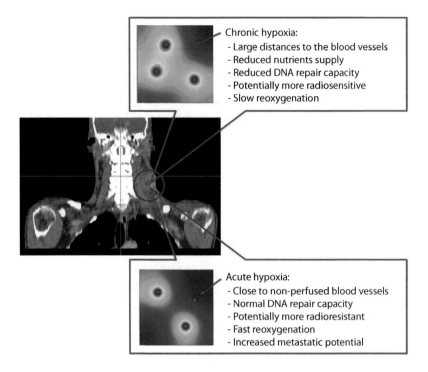

FIGURE 4.1 Illustration of formation of chronic and acute hypoxia in solid tumours in the head and neck area and the differences between them.

and acutely hypoxic cells might have implication going beyond the simple cellular survival as decreased DNA repair capacity in chronic hypoxia is expected to lead to increased mutation rates associated with chromosomal instability (Kumareswaran et al. 2012).

Molecular factors might further elucidate the differences between acute and chronic hypoxia observed in solid tumours including HNC. Among them, one could mention the differences in hypoxia-inducible factor 1α-mediated transcription, altered protein translation and differential activation of hypoxia-associated cell cycle checkpoints (Koritzinsky et al. 2005, 2006; Koritzinsky & Wouters 2007; Wouters et al. 2005; Rey 2016) which lead to different biological consequences.

One of the main characteristics of the tumour hypoxia is its dynamic character. The changes in the oxygenation of the tumour cells during the course of fractionated radiation therapy treatments are often referred to as reoxygenation, although, in the strict sense, the process is not unidirectional as hypoxic cells will change status to well-oxygenated cells, but bidirectional, as oxygenated cells could also become hypoxic. There are two types of reoxygenation which differ in terms of underlying mechanisms as well as time scale; slow and fast reoxygenation (Hall & Giaccia 2011). Slow reoxygenation occurs during the course of fractionated radiotherapy due to the preferential killing of the radiosensitive well-oxygenated cells close to the blood vessels rather than the radiation-resistant cells lacking oxygen. Consequently, as the treatment progresses, the number of cells near the blood vessels consuming oxygen decreases, resulting in larger oxygen diffusion distances, and the oxygen and nutrients become available to cells further from the blood vessels, which were previously deprived. This process is rather lengthy, the rate and the magnitude of slow reoxygenation being difficult to be predicted. Fast reoxygenation is related to perfusion processes and the re-opening of the blood vessels that were once occluded lead to local cellular oxygen changes and transient acute hypoxia. Thus, hypoxic-resistant cells at the time of delivering one fraction of radiation might become radiosensitive through fast reoxygenation in due time and could be killed by subsequent treatment fractions of radiation (Ljungkvist et al. 2005, 2006). The dynamics of tumour hypoxia in general and the reoxygenation of head and neck tumours treated with radiotherapy in particular have been investigated in several clinical studies based on PET imaging (Eschmann et al. 2007; Zips et al. 2012; Bittner & Grosu 2013) concluding that the presence of hypoxic-persistent subvolumes is inversely correlated with the response to radiotherapy and that improved tumour oxygenation is an indicator of positive response to treatment.

Therefore, experimental and clinical evidence indicates the presence of limited diffusion events and limited perfusion events as well as reoxygenation in solid tumours in general and in HNC in particular, which might result in a different response of the cells not only regarding repair and consequently cellular survival but also the metastatic spread, making the assessment of tumour oxygenation prior the start of the treatment imperative. For radiotherapy treatments, in particular, the identification of tumour hypoxia and its qualitative and quantitative description before the start of the treatment might be regarded as a window of opportunity for personalised treatment optimisation.

4.2. METHODS TO IDENTIFY TUMOUR HYPOXIA IN HNC

Due to its well-recognised role as global modulator of treatment response, the assessment of tumour oxygenation has been one of the clinical priorities since the early days of radiotherapy leading to the development of several methods of identifying tumour hypoxia in HNC. Generically, the modalities for determining the presence of hypoxia in tumours could be divided in two broad categories, depending on their invasive or non-invasive character.

Most of the early clinical evidence of the existence of hypoxia in HNC and its role as negative prognostic factor for the treatment outcome is based on in vivo measurements using *polarographic oxygen electrodes*. This technique, based on the insertion of a polarographic needle in the tumour, belongs to the invasive category. With a history of more than 100 years since Daneel showed, for the first time in 1897, that if a small potential is applied to platinum electrodes in aqueous media, the electrical current that appears is proportional to the oxygen tension in the media due to the electrolysis of dissolved oxygen (Hahn 1980), the polarographic method is one of the most widely used devices for measuring tumour oxygenation. Since then, various types of oxygen electrodes were developed and the oxygen electrode used in practice today, commonly known as the Eppendorf histograph, is manufactured by Eppendorf-Netheler-Hinz (Hamburg, Germany) (Vaupel et al. 1991). Several clinical studies have shown that median values of the oxygenation levels measured with the Eppendorf histograph in head and neck primary tumours and enlarged cervical lymph nodes range within radiobiological hypoxia, corresponding to partial oxygen pressure values lower than 10 mmHg (Becker et al. 2002; Nordsmark &

Overgaard 2004; Vaupel 1997). There are also several clinical studies that focused on the assessment of the tumour hypoxia in advanced-stage HNC using a polarographic electrode as a prognostic factor for locoregional-control, disease-free survival as well as overall survival (Nordsmark 1996, 1997, 1999; Stadler et al. 1999; Rudat et al. 2000, 2001; Dunst et al. 2003). The results of these studies are consistently showing the adverse prognostic character of low oxygen partial pressure values below 2.5 to 10 mmHg identified in either primary HNC or invaded lymph nodes for local tumour control, disease-free survival and overall survival among patients treated with radiotherapy alone or in combination with chemotherapy and surgery.

Due to the strong, positive correlation between the pre-treatment hypoxia measurements and the treatment outcome, the polarographic electrode measurements are considered to be the golden standard of in vivo measurements of tumour oxygenation in spite of the fact that the intrinsic principles of the technique do not allow the detection of the extreme values but only an average oxygen tension in few points within the tumour without providing spatial information regarding the location of the hypoxic regions in the tumour. This disadvantage comes in addition to the fact that the electrode actually consumes oxygen from the tissue. An alternative method, still invasive, proposed to overcome the problem of oxygen consumption, using a *time-resolved luminescence-based optical sensor*, OxyLite™ (Oxford Optronixy, Oxford, UK) is based on the oxygen-dependent quenching of the fluorescence of a luminophor deposited on the tip of a probe that can be inserted into the tissue (Collingridge et al. 1997; Young, Vojnovic & Wardman 1996). This method, however, suffers from the same limitation as the polarographic one with respect to the lack of information regarding the spatial distribution of the hypoxic regions within the tumour and the fact that it renders average oxygen concentration in large volumes.

The recent progress in imaging techniques has contributed not only to improving diagnosis and prognosis in HNC by early detection and increased accuracy of defining the targets but also by allowing in an almost non-invasive manner the characterisation of the tumour in terms of phenotypes known to be related to poor treatment outcome, such as tumour hypoxia. Thus, functional imaging allows the assessment of tumour hypoxia not only qualitatively but also quantitatively at the same time, with overcoming the major limitation of the polarographic and time-resolved oxygen probes, namely, the lack of spatial information regarding the location of tumour hypoxia (Bittner et al. 2013; Gregoire 2015). There are several imaging techniques and modalities clinically available today that could be used to gather information regarding the oxygenation of HNC, among which the magnetic resonance methods, such as *blood oxygen level-dependent magnetic resonance imaging* (BOLD-MRI) and *dynamic contrast-enhanced magnetic resonance imaging* (DCE-MRI) are the most promising ones in terms of spatial and temporal resolution (Bhatnagar et al. 2013; King 2016) although they do not allow direct measurements of oxygen partial pressure in tumours but only surrogates. BOLD-MRI is most likely to indicate acute hypoxia as it uses a paramagnetic contrast agent, deoxyhaemoglobin, to target the red blood cells in the blood vessels and hence gives information about perfusion-related events. Similar limitations could be listed for DCE-MRI, which is also mainly an indicator of tumour perfusion (Newbold et al. 2009; Agrawal et al. 2012).

Among the nuclear medicine imaging techniques, PET appears to be the most promising one, not only for assessing the tumour hypoxia before the treatment but also for following the changes in the tumour oxygenation as described in the recent comprehensive review by Marcu and colleagues (Marcu, Harriss-Phillips & Filip 2014). The nuclear medicine techniques build upon the pre-clinical extensive good experience not only in terms of characterising the severity but also to the extent of tumour hypoxia by using exogeneous bioreductive markers, chemical compounds that are metabolically reduced under hypoxic conditions, such as pimonidazole (1-[(2-hydroxy-3-piperdinyl)propyl]-2-nitroimidazole) and EF5 (2-(2-nitro-1H-imidazol-1-yl)-N-(2,2,3,3,3-pentafluoropropyl)acetamide). The resulting products are then observed by immunofluorescence, immunochemical, or flow cytometric methods (Chapman, Franko & Sharplin 1981; Evans et al. 1995; Hodgkiss, Webster & Wilson 1995; Varia et al. 1998; Raleigh et al. 2000; Evans et al. 2000, 2001). However, the use of the chemical compound such as the pimonidazole has the disadvantage of giving the unlikely indication of the cells being in an acutely hypoxic state or those that are energy-deprived to the point of biochemical incompetence since the metabolisation of the compound requires that the cell has to be in the hypoxic state long enough to allow the actual metabolisation process.

Several tracers are now available for in vivo imaging tumour hypoxia with PET being primarily divided in two main classes of hypoxia markers: 2-nitromidazoles and non-imidazoles (Apisarnthanarax & Chao 2005).

Fluoromisonidazole (FMISO) was the first nitroimidazole derivative radiolabelled with ^{18}F proposed for hypoxia imaging with PET and currently is the most used hypoxia tracer (Rasey et al. 1987; Lee & Scott 2007). An example of the FMISO uptake in two large hypoxic head and neck invaded lymph nodes is shown in Figure 4.2.

The relatively low uptake of FMISO in hypoxic lesions coupled with its slow clearance from the well oxygenated healthy tissue has led to the development of other ^{18}F-labelled nitroimidazoles like Fluoroetanidazole (FETA) and Fluoroazomycinarabinofuranoside (FAZA) (Piert et al. 2005; Krohn, Link & Mason 2008) and non-imidazole tracers like Cu(II)-diacetyl-bis-N-(4)-methylthiosemicarbazone (Cu-ATSM) (Lewis et al. 1999). More recently, the 2-nitroimidazole nucleoside analogue, 3-[^{18}F]fluoro-2-(4-((2-nitro-1H-imidazol-1-yl)methyl)-1H-1,2,3-triazol-1-yl)propan-1-ol ([^{18}F]HX4), was developed as a potential

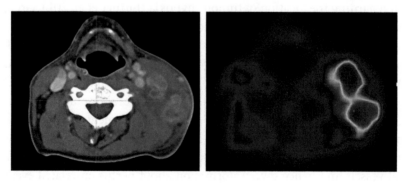

FIGURE 4.2   CT image and the corresponding FMISO PET image showing the high tracer uptake in two large hypoxic head and neck invaded lymph nodes.

marker and radiosensitiser for hypoxic tumour cells. Because of better water solubility and faster clearance, it is expected that [^{18}F]HX4 has better pharmacokinetic properties than currently used nitroimidazole hypoxia markers, such as [^{18}F]FMISO (Chen et al. 2012). Hypoxia imaging for HNC has been extensively investigated, showing that the uptake of hypoxic tracers is related to patient prognosis (Rajendran et al. 2006; Eschmann et al. 2005; Kikuchi et al. 2011; Rischin et al. 2005).

In spite of the clinically promising results of using PET for characterising tumour oxygenation, one has to keep in mind that this technique does not allow the absolute determination of the oxygen partial pressure.

A potentially promising technique that actually does allow the absolute determination of the oxygen partial pressure in vivo is electron paramagnetic resonance (EPR) oximetry. It was recently shown that EPR could be used to qualify or disqualify other oxygen imaging methods and that good correlation was found between EPR oxygen measurements and PET hypoxia imaging with the ^{18}F-labelled nitroimidazole derivative FAZA (Tran et al. 2012). The full potential of the technique could be further pursued for overcoming the most critical limitation of the current approach for assessing tumour hypoxia.

A special class of methods for identifying hypoxia in HNC is based on the expression of endogenous molecular markers of cellular hypoxia, such as hypoxia-inducible factor 1 (HIF-1) (Hill et al. 2015). HIF-1 is one of the most important transcription factors responsible not only for cell survival under radiobiological hypoxia but also for controlling angiogenesis, growth factors and vasoactive peptides, the metastatic potential, DNA replication, protein synthesis, pH regulation, and the cellular metabolism through promoting the genes controlling them, such as erythropoietin (EPO), the glucose transporters Glut1 and Glut3, and the vascular endothelial growth factor (VEGF) (Bristow & Hill 2008). One of the genes transcriptionally regulated by HIF-1α, carbonic anhydrase IX (CA9), might be of particular interest in HNC as it has been reported to be expressed in most carcinomas in this anatomical region and its expression appears to be not only spatially but also quantitatively correlated to the vessel density and the presence of necrosis (Beasley et al. 2001; Kaanders et al. 2002a; Hui et al. 2002; Koukourakis et al. 2002; Koukourakis et al. 2008; Schrijvers et al. 2008; Kwon et al. 2015). Experimental evidence also indicates the fact that HIF-1 expression is correlated with FMISO uptake in HNC (Sato et al. 2013).

A summary of the main techniques for determining the presence and quantifying tumour hypoxia in HNC together with their main applications is given in Table 4.1.

4.3 METHODS TO OVERCOME TUMOUR HYPOXIA

Preclinical and clinical evidence has established that low oxygen partial pressure in head and neck tumours is one of the main factors that determine the failure of radiation treatment (Overgaard 1994; Nordsmark; Overgaard & Overgaard 1996; Brizel et al. 1997; Overgaard 2007). Furthermore, it is also well known that the presence of hypoxia is not only a negative prognostic factor for the local failure of HNC radiotherapy but it has also been associated with tumour invasion and increased distant metastatic spread (Brizel et al. 1996). It is, therefore, very important to identify the presence of tumour hypoxia at the

TABLE 4.1   A Summary of the Main Techniques for Identifying and Quantifying Tumour Hypoxia in HNC

Technique	Main Principle	Measurement Outcome	Main Applications
Invasive Techniques			
Polarographic electrodes (Eppendorf histograph)	Electrolysis of dissolved oxygen giving rise to a flux of ionic species producing an electric current that has the intensity proportional to the concentration of the oxygen in the surrounding tissue.	Average values of pO_2 at the time of measurement in several measuring points in the tumour and the corresponding histograms. Clinical measurements are usually reported as hypoxic fraction (HF) of values lower than a given hypoxic threshold (2.5–10 mmHg).	Assessment of tumour hypoxia (HF) as prognostic factor for locoregional control, disease-free survival, and overall survival.
Time-resolved luminescence-based optical sensor (OxyLite™)	Oxygen dependent quenching of the fluorescence of the luminophor inserted into the tissue.	Spatial and continuous real-time tissue pO_2 measurements in several measuring points in the tumour.	Assessment of tumour hypoxia as a prognostic factor for poor treatment outcome.
Non-Invasive Techniques			
PET	Labelling with positron-emitting radionuclides exogeneous bioreductive markers, which are selectively accumulated in hypoxic tissue.	Spatial distribution of the measured activity expressed as standard uptake value (SUV).	Delineation of the hypoxic sub-volumes that should benefit from dose boosting based on the tumour to blood ratio (T/B) or tumour to muscle ratio (T/M) of tissue-specific radioactivity.
MRI	DCE-MRI is based on the diffusion of a MRI intravenous contrast agent (mostly gadolinium based) in the extracellular extravascular space (EES) of the tumour. BOLD-MRI is based on the paramagnetic effect of blood deoxyhaemoglobin.	The most common outcome parameter in DCE-MRI is the transfer function from plasma to EES, commonly referred to as volume transfer constant (K_{trans}). In BOLD-MRI, the reported parameter is the transverse relaxation rate of water in blood and surrounding tissue (R_2^*).	Assessment of the specific parameters (K_{trans} or R_2^*) as prognostic factors for poor treatment outcome.

planning stage of the treatment in an attempt to select the patients that might benefit from tailored treatment strategies for overcoming tumour hypoxia.

Given the well-recognised role of tumour hypoxia in the cure of HNC with radiotherapy, several approaches and techniques have been explored over the years for overcoming the reduced sensitivity to radiation associated with the lack of oxygen. However, in spite of the apparent plethora of modalities for overcoming tumour hypoxia, none of them has become part of the treatment routine in the management of HNC. The main modalities for overcoming tumour hypoxia could be divided into five main categories: (1) dose escalation based on hypoxia imaging, (2) the use of high LET radiation, (3) enhancement of oxygen delivery, (4) hypoxic cell radiosensitisers and (5) hypoxic cell eradication. The time point along the years of development of radiotherapy when they were proposed, the pace at which they were introduced into the clinics, and their clinical impact differ largely among the different modalities.

At present, the modality that appears to be explored most commonly is dose escalation based on *hypoxia imaging*. The general idea behind this approach is to take advantage of the fact that current molecular and functional imaging modalities allow not only the identification of tumour hypoxia but also the segmentation of the target into regions of different assumed radiosensitivity. Currently, the most used technique for imaging the regions that might benefit from dose escalation for overcoming their resistance associated with tumour hypoxia in HNC is PET. The prescription of the dose to the hypoxic sub-volumes is far from being standardised, several approaches being currently explored starting from empirical escalation up to the tolerance of the organs at risk and the normal tissues around the tumour (Chao et al. 2001; Grosu et al. 2007; Lee et al. 2009) to more sophisticated techniques matching the signal intensity in the PET or MR image with a given dose level, as in the so-called 'dose painting' approach (Alber et al. 2003; Flynn et al. 2008) or the heterogeneous dose prescription based not only on the level of the tracer uptake but also on tracer perfusion derived from the dynamic PET scan as proposed by Thorwarth and colleagues (Thorwarth et al. 2007). However, these approaches, which result in highly heterogeneous dose prescriptions, might fail to deliver the expected results because they do not take into account the dynamic character of hypoxia. Thus, in absence of frequent re-planning sessions based on repeated scans aiming at monitoring the location and the extent of the hypoxic sub-volumes, the careful match between the high-dose regions and the potentially resistant foci performed before the start of therapy might be lost at some point during the treatment, therefore leading to a loss in the probability of controlling the tumour. A practical method to include functional information into treatment planning, taking into account the concerns raised above has been proposed by Toma-Dasu and colleagues (Toma-Dasu et al. 2012) and tested with respect to feasibility on head and neck tumours. The method renders a dose-prescription level based on the signal intensities in the PET images that are converted into radiation sensitivities, also taking into account potential local changes in tumour hypoxia during the course of the treatment. However, in spite of the encouraging results of the studies on dose-escalation approaches based on functional imaging of tumour hypoxia and the feasibility of applying them in the treatment of HNC, none of them has yet been validated through a clinical trial and, therefore, their clinical implementation is still awaiting.

A step forward into the development of methods to overcome tumour hypoxia based on functional and molecular imaging would be to include the particular features of tumour hypoxia into the biological objectives for treatment planning. This method has the potential of designing tailor-made treatments adapted to the particular features of the individual patients. Hence, it has the potential to increase the rate of local control above the current levels by specifically addressing the factors that may worsen the response to treatment.

The dose-enhancement techniques mentioned above have mainly been developed for external bean photon therapy. However, one of the most quoted advantages of ions heavier than protons is the ability to overcome tumour hypoxia. Indeed, pre-clinical radiobiological experiments have shown that the oxygen enhancement ratio decreases with the increase in LET of radiation, approaching unity for very high LET values for a variety of cell lines, including human salivary gland (HSG) tumour cells (Furusawa et al. 2000). It is, therefore, not surprising that C-ions radiation therapy is considered a strong option for the management of HNC, which represents the third most common cancer indication treated with C-ions after prostate and lung at the National Institute of Radiological Sciences in Japan (Tsujii et al. 2008). However, in order to reach an oxygen enhancement ration of 1, the LET of radiation has to be very high, much higher than the LET in a typical spread-out Bragg-peak (SOBP). Thus, the classic paradigm of C-ion treatments overcoming tumour hypoxia has recently started to be questioned since the dose-averaged LET in the SOBP that covers a typical tumour might be much lower than the LET in the pristine Bragg peaks used for irradiation of cells in in vitro clonogenic cell survival experiments, which established the pre-clinical evidence (Antonovic et al. 2014, 2015).

Another class of methods to overcome tumour hypoxia is based on enhancing oxygen delivery to the tumour cells. This could be achieved through *hyperbaric oxygen therapy*. The technique consists of placing the patient in a hyperbaric oxygen chamber before the start of the radiotherapy treatment. The hyperbaric oxygen chamber is a sealed chamber that has been pressurised at 1.5–3 times normal atmospheric pressure in which the patient is breathing pure oxygen (Mayer et al. 2005). Thus, the main idea behind this approach is to overcome hypoxia-induced radiation resistance by increasing vascular partial pressure of oxygen. So far, promising results have been achieved in HNC treatments (Henk 1986; Ferguson, Hudson & Farmer 1987; Haffty, Hurley & Peters 1999) but the difficulties around the logistics for using the technique have hindered its wide clinical implementation.

An alternative to breathing pure oxygen in the hyperbaric oxygen chamber is carbogen, a hyperoxic gas mixture consisting of 95%–98% O_2 and 2%–5% CO_2. Breathing carbogen will, therefore, have a similar effect to enhancing the oxygen content and thus allowing diffusion to larger distances from the blood vessels, which would counteract chronic hypoxia. Carbogen is one of the main factors behind the Accelerated Radiotherapy with CarbOgen and Nicotinamide (ARCON) protocol applied in the management of HNC. Accelerated therapy in the context of the ARCON protocol refers to a reduced overall treatment time from about 7 weeks to around 5 weeks while the number of fractions and the total dose remain the same. ARCON combines accelerated fractionated radiotherapy in a complex manner to counteract cellular repopulation, with the use of the hyperoxic gas carbogen to reduce diffusion-limited hypoxia and nicotinamide, a vasoactive agent, to reduce perfusion-limited hypoxia. The ARCON protocol has been tested in a phase II clinical trial on 215 patients with advanced head and neck squamous cell carcinoma (HNSCC), resulting in the local control rates of 80% for larynx, 60% for hypopharynx, 87% for oropharynx, and 29% for oral cavity in patients with T3 and T4 tumours (Kaanders et al. 2002b), showing thus promising results and prompting further phase III randomised studies. In the phase III ARCON trial, 345 patients were accrued. The local tumour control rate at 5 years was 78% for the accelerated radiotherapy arm (AR) versus 79% for ARCON after a median follow-up of 44 months. The 5-year regional control of 93% for ARCON was considerably higher than the corresponding one for the AR arm, which was only 86%. It should be noted that the improvement in regional control was particularly observed in patients with hypoxic tumours (100%) and not in patients with well-oxygenated tumours (55%) (Janssens et al. 2012). Furthermore, the results of a subsequent study based on the same cohort of patients included in the phase II ARCON trial showed that ARCON protocol has the potential to correct the poor outcome of cancer patients with anaemia (Janssens et al. 2014).

The negative influence on radiotherapy outcome of the lack of oxygen in HNC could also be counteracted by the use of *hypoxic cell radiosensitisers* that mimic the effect of oxygen in the radiochemical process of DNA damage induction and fixation. The most used hypoxic cell radiosensitisers are nimorazole and misonidazole (Cottrill et al. 1998; Overgaard et al. 1998a, 1998b). The effect of nimorazole was investigated in the large Danish Head and Neck Cancer Study (DAHANCA) Protocol 5-85 phase III randomised trial (Overgaard et al. 1998b). The results showed that the use of nimorazole as radiosensitiser significantly improves the effect of radiotherapy of supraglottic and pharyngeal tumours with only mild toxicity. The role of nimorazole was also assessed in combination with another complex protocol, Continuous Hyperfractionated Accelerated Radiation Therapy (CHART) in locally advanced stage IV HNSCC indicating better tumour response in comparison with CHART regimen alone (Cottrill et al. 1998). The findings of this study were further confirmed by the one trial conducted on a larger number of patients with advanced stage III or IV HNSCC (Henk, Bishop & Shepherd 2003).

Another modality for overcoming tumour hypoxia is the use of *bioreductive drugs*, compounds that are reduced intracellulary to form cytotoxic agents such as mitomycin C and tirapazamine (Overgaard 2007). Tirapazamine exhibits cytotoxicity in hypoxia cells but not in normal cells which makes it a suitable candidate for future trials in HNC radiotherapy. To date, the clinical studies, including tirapazamine as hypoxic cytotoxic agent, rendered various results, depending on the patient selection criteria (Le et al. 2012) and, although promising, the role of tirapazamine in the management of HNC should be further investigated.

A special class of methods for overcoming tumour hypoxia relies on the actual destruction of hypoxic cells by using *hyperthermia*. Thus, by transmitting radio waves in the tissue, the actual heating of the cells, or hyperthermia, is generated. After a short thermal shock, proteins emerge on the tumour cell surface. As a consequence, cells of the immune system are activated and can efficiently destroy tumour cells, which carry such thermal shock molecules. At the same time, hyperthermia can be used as a radiosensitiser. Significant clinical benefits from combined hyperthermia and radiotherapy have been reported (Datta et al. 2016).

Finally, one could mention a rather unorthodox approach to handling tumour hypoxia. Instead of overcoming it, one could exploit hypoxia by targeting the tumour-specific genes activated in the presence of tumour hypoxia, thus taking advantage of the fact that very little hypoxia is observed in normal tissues (Rey et al. 2017).

TABLE 4.2   A Summary of the Main Modalities for Overcoming Hypoxia-Induced Radio Resistance in the Management of HNC

Dose Escalation
• Empirical dose escalation based on functional imaging of tumour hypoxia
• 'Dose painting' based on based on functional imaging of tumour hypoxia
High LET Radiation
• C-ions radiotherapy
Enhancement of Oxygen Delivery
• Hyperbaric oxygen therapy
• Accelerated Radiotherapy with CarbOgen and Nicotinamide (ARCON) protocol
Hypoxic Cell Radiosensitisers
• Nimorazole and misonidazole
Hypoxic Cells Eradication
• Cytotoxic bioreductive drugs (e.g. tirapazamine)
• Hyperthermia

A summary of the main modalities, including overcoming the radiation resistance associated with tumour hypoxia, is given in Table 4.2.

4.4 ANGIOGENESIS AND VEGF

As it has previously been mentioned in this chapter, tumour hypoxia not only has a negative effect on local control but also promotes the invasive character and the formation of distant metastases. One of the main reasons often cited is the stimulation of the angiogenic growth factors in general and the VEGF in particular. This observation has opened a special window of opportunity in the management of advanced HNC, namely the anti-angiogenic treatment targeting the VEGFs. In HNCa in particular, the anti-angiogenic treatments that have shown promising results involve bevacizumab, which are often combined with cisplatin-based chemo/radiotherapy (Jain 2014).

CHAPTER 5

The Mechanisms Behind Tumour Repopulation

5.1 INTRODUCTION

One of the Rs of radiotherapy refers to regeneration or repopulation of cells during treatment (Withers 1975). While dose fractionation in radiotherapy allows normal cells to repopulate the affected tissue or organ over subsequent fractions, this process is also valid for the surviving tumour cells, which can regenerate the tumour during treatment. This behaviour can lead to a poor outcome, a reason why repopulation of cancer cells is considered another culprit for treatment failure (Marcu & Yeoh 2012).

Radiotherapy causes cell killing, which allows the surviving tumour clonogens the opportunity to multiply with a reduced cell-loss factor and possibly shorter cell cycle. This behaviour was shown to be characteristic to squamous cell carcinomas and was labelled accelerated repopulation (Withers, Taylor & Maciejewski 1988). Experimental data have shown that the accelerated repopulation of cells belonging to squamous epithelia (Dörr 1997) is characteristic of the acute response of hierarchical normal tissues after injury (Figure 5.1). Trott and Kummermehr (1991) suggested that squamous cell carcinomas preserve some homeostatic control mechanisms specific of their tissue of origin. Other studies also provided evidence whereby normal and cancerous squamous epithelial cells exhibit a very similar behaviour in response to cell loss or other injury (Denham et al. 1996; Trott 1999). Revolutionary work showed that apoptotic tumour cells might be responsible for tumour repopulation during cytotoxic therapy. Huang et al. (2011) demonstrated, both in vitro and in vivo, that apoptotic cells encourage the repopulation of irradiated tumour cells more proficiently than non-irradiated cells. Their investigation showed that caspase 3, a cysteine protease that is involved in the execution of apoptosis, is the driving force behind repopulation, by regulating growth-promoting signals that are generated from dying cells. This observation was confirmed by HNC patient studies, where high levels of caspase 3 expression were associated with a higher rate of tumour recurrence, due to the

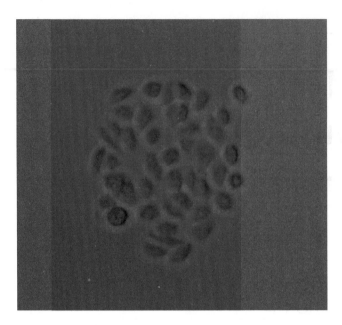

FIGURE 5.1 Colony forming ability of HNSCC cells originating from an UM-SCC-1 cell line. (Courtesy of Paul Reid & Eva Bezak, University of South Australia.)

compensatory proliferation that is stimulated by apoptotic cells. More recent studies confirmed the role of caspase 3 in stimulating proliferation of tumour cells in other aggressive cancers, such a pancreatic ductal adenocarcinoma (Cheng et al. 2015).

Radiation is not the only agent that can trigger such a response, given that substantial cell kill occurs also after chemotherapy. While the information about repopulation during chemotherapy is limited, the available data suggest that the rate of repopulation may increase with subsequent cycles of chemotherapy (Davis & Tannock 2000). Furthermore, Peters and Withers (1997) indicated that surgery can also trigger repopulation in operable HNCs starting from those tumour foci that are unintentionally left behind.

Radiobiological studies investigating accelerated repopulation of human squamous mucosal cells revealed that cell proliferation already begins 1 week after the start of treatment (Dörr et al. 2002). There are several mechanisms considered to be responsible for this accelerated repopulation, and they are: cell recruitment, accelerated cell division and/or loss of asymmetrical division of CSCs, and abortive division (Table 5.1). While accelerated stem division and abortive division are thought to be controlled by tissue hypoplasia, the process of asymmetry loss is dictated by stem cell depletion (Dörr 2003).

As defined in Table 5.1, abortive division is characteristic to differentiated cells, which can undergo a finite number of cell divisions before they become sterile (Dörr 2009). Although they have a finite lifespan, these cells are still contributing, to a certain extent, toward cell growth during treatment. Nevertheless, as also shown by Monte Carlo modelling, repopulation via abortive division neither requires alterations of the treatment schedule nor dose escalation for tumour control (Marcu 2014). Therefore, only the first three

TABLE 5.1   Possible Mechanisms Behind Repopulation in Head and Neck Tumours

Repopulation Mechanism	Description	Cell Type Involved
Recruitment	The process whereby quiescent cells re-enter the mitotic cycle when triggered by various stimuli (i.e., cell loss).	CSC
Loss of asymmetrical division	The more common asymmetrical division that creates one stem cell and a non-stem cancer cell is replaced by symmetrical division that creates two CSCs in mitosis.	CSC
Accelerated cell division	The shortening of normal cell cycle duration, which leads to a higher cell turnover than before repopulation was triggered.	CSC
Abortive division	A limited division capacity of differentiated cells, creating a few generations of non-stem cancer cells. While a counterbalance of lost cells occurs, this mechanism does not lead to unrestrained repopulation.	Differentiated cell

mechanisms are of interest from a tumour repopulation perspective and will be described in more detail in the later sections.

5.2 CANCER STEM CELLS

5.2.1 The Concept of Cancer Stem Cell

Cancer stem cells are not a new concept in the scientific literature. Decades ago they were referred to as clonogenic cells, until it became clear that these cells display stem-like properties (Baumann, Krause & Hill 2008), similar to normal stem cells. The CSC model stipulates that malignant tumours are organised hierarchically and their growth is driven by a small subgroup of CSCs that can also undergo epigenetic transformation, the equivalent of cellular differentiation of normal cells (Shackleton et al. 2009).

Therefore, CSCs represent a subpopulation of tumour cells that exhibit a number of characteristics that distinguish them from other cancer cell types: are able to proliferate indefinitely, can create all lineages of the original tumour, are more quiescent than non-CSCs, and can regenerate through both symmetric and asymmetric division (Table 5.2). The definition of CSCs is now widely accepted by the scientific community and to illustrate their pluripotency, these cells are also referred to as 'tumour-initiating cells'.

5.2.2 Cancer Stem Cells in HNC

Experimental evidence among tumour cell lines shows that the percentage of CSC varies greatly within tumours of different histopathological type, from as low as 0.4% in acute myeloid leukaemia, to as high as 82.7% in colon cancers, although these values might strongly depend on the specificity of the stem cell markers used (Huang et al. 2013). In HNC, the first qualitative report on the identification of CSC is likely that of Prince et al. (2007) who has isolated a subpopulation of cells with stem-like properties.

Even among HNC cell lines, there are significant differences when it comes to CSC quantification. Consequently, Tang et al. (2013) showed that the CSC proportion in various head

TABLE 5.2   The Main Properties of Cancer Stem Cells as Shown by Experimental Evidence

Properties of Cancer Stem Cells	References
Can initiate cancer growth	Reya et al. (2001)
Are able to generate all heterogeneous lineages of the original tumour	Al Hajj et al. (2003)
Have the potential to recreate themselves by symmetrical division	Morrison & Kimble (2006)
Can be recruited from their niche into the proliferating pool	Vlashi, McBride & Pajonk (2009)
Have the ability to proliferate indefinitely (self-renewal)	Reya et al. (2001); Moore & Lyle (2011)
Demonstrate higher resistance to treatment than non-stem cancer cells	Moore & Lyle (2011)
Display enhanced DNA repair ability	Moore & Lyle (2011)
Are able to trigger tumour repopulation as a response to treatment-induced cell kill	Koukourakis et al. (2012)
Can be active and invasive (migratory CSCs) or quiescent and non-invasive (stationary CSCs)	Geissler et al. (2012)
Preferentially reside in specific microenvironmental niches within the tumour (HNCs reside in perivascular niches)	Krishnamurthy et al. (2010)
Migratory CSCs overexpress genes related to invasiveness, which can contribute to metastatic spread	Shrivastava et al. (2015)
Display cellular plasticity that enables changing from CSC to non-CSC state and vice versa	Cabrera, Hollingsworth & Hurt (2015)

and neck cell lines ranged between 1.7% and 13.5%, whereas the investigations undertaken by Harper et al. (2007) on CaLH3 cell lines have indicated a proportion of CSC of 12.3%. The variation in CSC proportions reported in the literature is strongly dependent on the markers used to identify CSCs and their interaction. Cell surface markers, such as CD (cluster of differentiation) molecules, are currently used for the identification of specific cells, such as CSCs. CD markers are surface proteins that enable the study of cell differentiation and they belong to different classes: adhesion molecules, integrins, glycoproteins, or receptors.

A CSC-specific marker involved in malignancies of epithelial origin is the hyaluronic acid receptor CD44, and was more often found to be overexpressed in a number of cancers than any other surface marker (Zöller 2011). In HNC, the role of this marker as a predictive factor was investigated by several research groups (Baumann & Krause 2010; Joshua et al. 2012; Trapasso & Allegra 2012). While the prognostic value of CD44 is still controversial, it was found that overexpression in pharyngeal and laryngeal cancer is related to worse T and N staging and prognosis; however, no clear correlation was found between CD44+ and oral cancer outcome (Chen et al. 2014). In oropharyngeal cancers, the results are also inconsistent, as the correlation between CD44 expression and poor prognosis reported by some studies (Lindquist et al. 2012) could not be demonstrated by others (Rajarajan et al. 2012). In other anatomical sites related to the head and neck area, high expressions of CD44 were associated with greater invasiveness and distant metastases, as well as treatment resistance (de Jong et al. 2010; Yüce et al. 2011). A recent report based on a multicentre, retrospective study of the German Cancer Consortium Radiation Oncology Group that included 158 patients with locally advanced HNSCC revealed that the inclusion of CD44 protein expression in the data analysis has improved the prediction power considerably for the 2-year locoregional control (Linge et al. 2016).

While further investigations are required to consolidate the role of CD44 among CSC markers in HNC, the potential predictive value demonstrated so far suggests that the identification and targeting of CSCs could greatly improve treatment outcome.

Other CSC markers trialed in HNC are CD133, ALDH1 and ABCG2 (Clay et al. 2010; Chen et al. 2011; Yanamoto et al. 2014). CD133 is a glycoprotein first identified as a hematopoietic stem cell marker; later, it was being investigated as a surface marker in several solid cancers, including head and neck (Chen et al. 2011). While the molecular mechanisms mediated by this marker in regulating CSCs are not fully elucidated, it was shown that overexpression of CD133 in HNC is negatively correlated with survival and, moreover, it serves as an indicator of tumour repopulation and malignant progression (Chen et al. 2011). Cells that express the intracellular enzyme aldehyde dehydrogenase (ALDH) were found to be a subset of the CD44+ cells, given that cells exhibiting increased levels of ALDH were identified to also have elevated CD44 expression (Clay et al. 2010). A study that was investigating the tumorigenic ability of cells expressing high levels of ALDH in HNC concluded that the activity of ALDH on its own is a highly selective marker for CSCs (Clay et al. 2010). In vivo head and neck studies showed that ALDH+ cells comprise a population that preserves its tumorigenic abilities after irradiation and could trigger tumour regrowth (Kurth et al. 2015). The same study indicated that ALDH activity in HNC is mainly attributed to the ALDH1A3 isoform, given that the inhibition of ALDH1A3 expression led to reduced tumour radioresistance (Kurth et al. 2015).

ABCG2 is a member of the ABC transporter protein family that causes drug-resistance in cancer and was also shown to serve as a marker for stem cells in various cancers (Zhou et al. 2001). In a study of neoadjuvant chemotherapy for tongue cancer, Yanamoto et al. (2014) observed that cells expressing ABCG2 are resistant to chemotherapy; this expression being correlated with tumour progression and lymph node metastasis. Based on their results, the authors advise against neoadjuvant chemotherapy in patients with oral tongue squamous cell carcinoma.

While the fraction of CSCs may play an important role in tumour response to treatment, the absolute number is probably not the only CSC-related factor that influences the outcome. An interesting experimental observation revealed the fact that human papilloma virus (HPV)-positive HNCs can have elevated levels of CSCs, which is a counterintuitive result considering the good responsiveness of HPV+ HNC to treatment (Zhang et al. 2014). In an experiment undertaken on three HNC cell lines, Zhang et al. (2014) observed that the HPV+ cell line had about 3 times the number of CSCs compared to the negative cell line. This result might suggest that HPV+ CSCs may be phenotypically distinct from HPV– CSC and thus the cell phenotype is more critical in terms of treatment outcome than the absolute number of CSCs. Similar results were observed by Reid (2017) during flow cytometric studies of an HPV+ cell line (UM-SCC-47) and an HPV– head and neck cell line (UM-SCC-1). As illustrated in Figure 5.2, UM-SCC-47 cell cultures showed a mean population of CD44+/ALDH+ cells of 2.87 ± 0.219, compared with 0.57% ± 0.077 of the UM-SCC-1 population.

More research is needed in this area, as the above results could not be confirmed by other studies (Tang et al. 2012; Vlashi et al. 2016). Just the opposite, Vlashi et al. (2016) showed

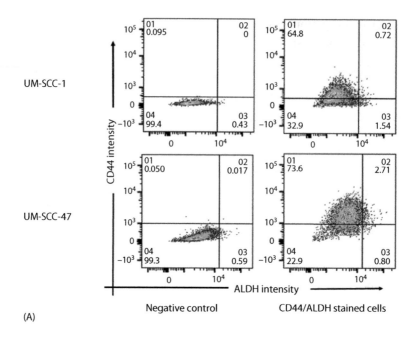

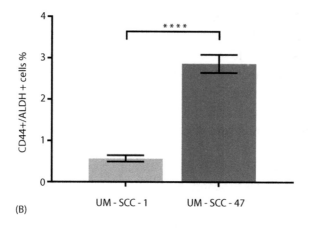

FIGURE 5.2 **(A)** Scatter plots from flow cytometry analysis of baseline proportions of cell phenotypes by CD44/ALDH expression. Upper right quadrants show percentages of cells positive for both CD44 and ALDH. **(B)** Comparison of CSC percentage in untreated cell lines. UM-SCC-47 shows a significantly higher proportion of CSC by CD44+/ALDH+ phenotype than UM-SCC-1. (Courtesy of Paul Reid, MSc Thesis 2017, University of South Australia.)

that HPV+ HNC actually has a lower number of CSCs; however, the difference in treatment response as a function of HPV-status is given by the distinction in radiation-induced dedifferentiation. This translates into the fact that non-stem cancer cells belonging to HPV− cell lines present with enhanced dedifferentiation ability stimulated by radiation compared to HPV+ cell lines (Vlashi et al. 2016). Through dedifferentiation, non-stem cancer cells gain stem-like properties, thus increasing the CSC population, which is responsible for poor outcome.

For more information on CSCs and their role in head and neck tumour development, the reader is directed to a recently published, comprehensive review of the literature compiled by Reid et al. (2017).

5.3 THE MECHANISMS BEHIND TUMOUR REPOPULATION

5.3.1 Cell Recruitment

While previously it was thought that cells become quiescent due to challenges imposed by nutrient deprivation, now it is known that a quiescent state allows the preservation of the cells' key features, the reason why stem cells choose to reside in a resting phase to preserve their genomic integrity over a long time period (Cheung 2013).

Cell recruitment, or the re-entry into the cell cycle, is one of the most natural repopulation mechanisms, considering the available pool of quiescent cells that reside in the G_0 phase (Figure 5.3). In response to various stimuli, such as irradiation or chemotherapy-caused cell loss, resting cells can re-enter the cycle and proliferate. Despite the commonness of this mechanism, the literature lacks of quantitative data regarding the occurrence of recruitment. Observations of untreated tumour population of hemopoietic cells, revealed that the proportion of recycled stem cells is below 2% of the total number of cells (Tubiana 1988). This value is, however, expected to be larger among tumours undergoing therapy, given that recruitment is stimulated by cell depletion.

It was demonstrated that the activation of *Notch* signalling that influences cell fate can stimulate quiescent stem cells to re-enter the cell cycle (Campa et al. 2008). While this process was evidenced in cardiomyocytes, activated Notch has been seen in several tumours. In HNC, the investigations on the molecular alterations of Notch signalling pathway are limited. More recent reports show that *Notch 1* mutations occur in 10%–15% of HN squamous cell carcinomas (Sun et al. 2014). Given that the Notch signalling pathway is linked to cell cycle exit and also to stem cell renewal ability, the role of Notch activation in HNC could play an important role in tumour repopulation through cell recruitment.

Furthermore, Phillips, McBride and Pajonk (2006) showed that fractionated doses of radiation can increase Notch 1 activation. The study that was undertaken on MCF-7 and MDA-MB-231 breast cancer cell lines showed that activated Notch 1 levels increased after fractionated radiotherapy (5 × 3 Gy) but not after a single dose treatment of 10 Gy.

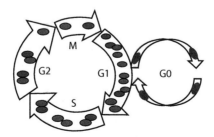

FIGURE 5.3 A sketch illustrating the tumour cell population in various phases of the cycle and the reversibility of the G_0 state.

Fractionated irradiation was demonstrated to promote self-renewal of CSCs and recruitment, increasing the cycling CSC population.

It was shown that both stem and differentiated cells can reside in the G0 phase, thus being capable of re-entering the cell cycle when stimulated (Cheung & Rando 2013). However, as far as tumour repopulation is concerned, due to their limited proliferative ability, differentiated cells will not contribute to regrowth, whereas CSCs will. More research is needed to quantify the proportion of quiescent CSCs that can potentially interfere with tumour control in HNC patients.

5.3.2 Accelerated Stem Cell Division

Accelerated stem-cell division is a biological process that results in the shortening of the stem cell cycle, which, in turn, leads to a higher rate of mitosis. Several investigators have identified this route of action as being correlated with accelerated repopulation (Withers & Elkind 1969; Hansen et al. 1996; Dörr 1997). Investigating the cell kinetics of the intestinal epithelium in normal versus irradiated animals, Lesher and Baumann (1969) have observed a shortening of the cell cycle time, using the percent labelled mitoses technique. Compared with the initial cell cycle time of 13.1 h after the delivery of 3 Gy whole body dose, they noticed a reduction of the gastrointestinal cell cycle time to 10.4 h, which has returned to normal within a week. After a higher dose of irradiation of 10 Gy, the shortening of the cell-cycle time was more remarkable, down to 7 h. This drastic measure came as a response to cell loss in order to rapidly repopulate the small intestine. Similar experiments have revealed a dose-dependent effect of repopulation, showing that for doses higher than 10 Gy when excessive cell loss occurs, repopulation could not be triggered (Hagemann, Sigdestad & Lesher 1972).

These early histological studies of jejunum demonstrate the effect of radiation in stimulating repopulation via shortening of the cell cycle. While undertaken on healthy cells, the study results could be translated to tumours, given that the crypt cells of the intestinal epithelium exhibit stem-like properties also found among tumour cells.

A Monte Carlo model of head and neck tumour growth and behaviour during radiotherapy showed that accelerated stem cell division is a powerful repopulation process; as for very short stem-cell cycle times, not even an accelerated fractionation regimen would qualify as more successful for locoregional control than standard radiotherapy (Marcu & Bezak 2012).

5.3.3 Loss of Asymmetrical Division of Cancer Stem Cells

Two of the main properties of stem cells are: (1) self-renewal, by creating identical copies of themselves and (2) lineage-specific differentiation, by creating non-stem, differentiated progenies.

Asymmetrical division of stem cells is the ability of the cell to produce a clone or copy of itself (i.e. another stem cell) and another non-stem daughter that is committed to differentiation (Knoblich 2008). Asymmetrical division is, therefore, required to produce cellular diversity and to maintain a balanced cell production. A disruption along the asymmetrical division path can result in abnormal growth (Gómez-López, Lerner & Petritsch 2014).

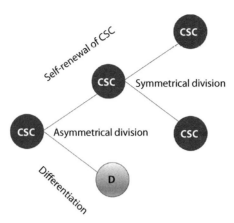

FIGURE 5.4   Asymmetrical versus symmetrical division of CSCs. The self-renewal ability of CSCs is illustrated as well as the generation of a differentiated (D) cell through asymmetrical division.

A proof of this is the fact that decreased asymmetrical division frequency was found in several cancer types, such as brain, breast, and leukaemia.

Nevertheless, CSCs retain their asymmetrical division ability, but symmetrical division starts to become more relevant and play an important role in tumour repopulation. The two division patterns coexist, although in various proportions, depending on the tumour (Cicalese et al. 2009). Through symmetrical division, CSCs generate two stem-like cells (Figure 5.4); a process that increases the pool of CSCs within a tumour. Cicalese et al. (2009) have shown that the tumour suppressor gene p53 regulates polarity of cell-renewing division and that the loss of this gene supports symmetric division of CSCs. They concluded that one of the physiological roles of the p53 gene is to uphold a constant number of stem cells by dictating an asymmetric division pattern for self-renewal. Hence, a loss of p53 was shown to increase the frequency of symmetrical divisions, stimulating cancer growth. The restoration of p53 could, therefore, affect CSC division and could be considered a step toward tumour dormancy.

Experimental as well as Monte Carlo models have revealed that the loss of asymmetrical division of CSCs is probably the main mechanism responsible for tumour repopulation during treatment (Trott 1999; Marcu, van Doorn & Olver 2004; Morrison & Kimble 2006).

5.4 CLINICAL IMPLICATIONS

5.4.1 Predictive Assays and Detection Methods

Several decades ago, the scientific community attempted to develop various predictive assays for pre-treatment evaluation of hypoxia, proliferation, and radioresistance, which largely failed when tested in vivo.

The limited success of cell kinetic assays in evaluating the proliferative potential of tumours required the development of new tools that would offer such information, preferably in a non-invasive way (Marcu, Bezak & Filip 2012). Positron emission tomography (PET) imaging with proliferation kinetics-specific markers is today the most accepted technique to evaluate the proliferative ability of tumours and to correlate it with treatment outcome. Several tumour proliferation markers labelled with radioisotopes have been tested as candidates for PET imaging.

The most accurate method for evaluating cellular proliferation is by measuring the extent of DNA-synthesising cells, through incubation with compounds that are able to be incorporated into the DNA. Thymidine analogues are ideal candidates to fulfill this criterion. A key enzyme in DNA synthesis is thymidine kinase 1 (TK-1), which has a peak activity during the S phase of the cell cycle. The tumour proliferation marker, 3'-Deoxy-3'[18F]-fluorothymidine (FLT) was shown to be phosphorylated by TK-1, making the radiolabelled FLT a possible candidate for the imaging of cellular proliferation. Given that TK-1 activity is elevated during the S phase of the cell cycle, ^{18}F-FLT uptake is well correlated with tumour proliferation. This property serves in assessing tumour response to chemotherapy, as persistent ^{18}F-FLT uptake after a drug cycle would translate either into drug ineffectiveness or limitations in drug perfusion (Bollineni et al. 2016).

Menda et al. (2009) reported the results of a kinetic analysis in HNC patients involving ^{18}F-FLT before and 5 days after 10 Gy radiotherapy and one cycle of chemotherapy. They observed that the high initial ^{18}F-FLT tumour uptake showed a significant reduction after treatment, correlated with decrease in thymidine kinase activity after treatment. In another study involving ten patients with oropharyngeal carcinoma, Troost et al. (2010) employed ^{18}F-FLT to identify tumour sub-volumes with high proliferation. Patients underwent serial ^{18}F-FLT-PET scans at baseline and during radiotherapy which helped assessing early response, tumour heterogeneity and sub-volumes of high proliferative activity within the gross tumour volume. The decrease in ^{18}F-FLT uptake was greater than twofold in the initial treatment phase and continued to decrease in the same rhythm by the fourth week of treatment. In correlation with the ^{18}F-FLT uptake, a significant decrease in gross tumour volume was generally observed, while the identification of residual ^{18}F-FLT sub-volumes enabled dose escalation to improve locoregional control (Troost et al. 2010). While ^{18}F-FDG is still the most widespread agent in PET imaging, latest studies have shown that pretreatment ^{18}F-FLT is a superior prognostic indicator compared to ^{18}F-FDG in HNC patients (Hoshikawa et al. 2015).

Another potential group of tumour proliferation-specific radiopharmaceuticals are radiolabelled sigma-2 receptor ligands. Sigma receptors are complex proteins that are involved in several aspects of cancer pathology. In vitro studies showed that the over-expression of sigma receptors in proliferating cells can be up to 10 times higher than in quiescent cells (Wheeler et al. 2000), which is a reason why radiolabelled ligands that bind to sigma-2 receptors are currently tested for PET imaging to provide a quantitative assessment of cell proliferation. A number of studies indicated that PET imaging with radiolabelled sigma-2 receptor ligands can offer superior tumour specific information compared to TK-1 based radiotracer imaging (Rowland et al. 2006; Mach, Dehdashti & Wheeler 2009). Carbon-11 as well as ^{18}F-labelled sigma receptor ligands have been investigated with different results (Mach, Dehdashti & Wheeler 2009). A disadvantage of ^{11}C-labelled compounds is the short half-life of ^{11}C (20.38 min), which limits its clinical utility, despite the high tumour uptake and good image contrast showed by in vivo studies (Mach, Dehdashti & Wheeler 2009).

A radioisotope with high affinity for sigma-2 receptors and clinically adequate half-life (16.2h) that was tested in vivo is bromine-76. Rowland et al. (2006) compared the potential of [76]Br-labelled ligands with the more established ^{18}F-FLT, to evaluate its suitability for proliferation-specific imaging in terms of tumour uptake. It was observed that 2 h post injection of the [76]Br-labelled agent, tumour to normal tissue ratio was about 9 times higher compared to FLT, which resulted in better tumour visualisation. Furthermore, the metabolic clearance of non-specifically bound [76]Br radioactive compounds was faster than ^{18}F-FLT clearance. While these results have justified further investigations, there are very limited studies investigating the possible implementation of [76]Br as a PET agent.

The marker protein Ki-67, an antigen present in the nuclei of cycling cells-only, is associated with cell proliferation and, therefore, is an indicator of proliferation rate within a tumour. A promising ^{18}F-labelled sigma-2-receptor ligand for PET called ^{18}F-ISO-1 was synthesised and trialed with success in animal (Tu et al. 2007) and human tumours, including head and neck (Dehdashti et al. 2013). The first human study aimed to investigate the dosimetry and safety of the compound and to correlate tumour uptake with Ki-67 tissue assays and mitotic index (Dehdashti et al. 2013). The results of this pilot study indicated that due to the significant correlation obtained between ^{18}F-ISO-1 uptake and Ki-67, the radiotracer can stratify patients into groups of high proliferative status (defined by Ki-67 >35%) and low proliferative status. The Ki-67 threshold value established by the trial can assist in the prediction of tumour response to standard chemotherapeutic agents that target the cycling population.

5.4.2 Altered Fractionation Schedules

In clinical terms, accelerated repopulation is a drastic increase in tumour growth rate after the commencement of radiotherapy. As shown by Withers (1993), this cell repopulation occurs with a doubling rate that is 15–20 times faster than in unperturbed tumours. Usually, repopulation becomes apparent 3–4 weeks after the treatment initiation, as a response to cell depletion. However, compensatory proliferation may also occur because of growth-depriving factors such as malnutrition and hypoxia (Turesson et al. 2003).

In rapidly proliferating tumours, such as head and neck carcinomas, the different response between tumours (characterised by fast proliferation and repopulation) and late responding normal tissue (characterised by slow proliferation and weak or no repopulation) works against treatment protraction and justifies shorter treatment time, with accelerated dose delivery. A radiotherapy schedule fitting these conditions is accelerated fractionation, that delivers an intensified dose to the tumour, exceeding the conventional 2 Gy a day, while also shortening the overall treatment duration by a couple of weeks (Table 5.3). Another treatment regimen designed to overcome HNC resistance during treatment is hyperfractionated radiotherapy that was successfully delivered in several clinical trials and showed improved outcome as compared to standard fractionation or even accelerated radiotherapy (Bourhis et al. 2006).

For many years, the standard of care in HNC treatment is the combined chemoradiotherapy, dominated by the cisplatin-based combined approaches. Altered fractionation radiotherapy with concomitant chemotherapy was shown to increase locoregional tumour control due to drug-radiation interaction.

TABLE 5.3   Treatment-Specific Parameters for Standard and Altered Fractionation Schedules

Fractionation/Treatment Parameters	Standard	Accelerated	Hyperfractionated	Accelerated Hyperfractionated
Total dose	70 Gy	<70 Gy	≥70 Gy	<70 Gy
Dose/fraction	2 Gy	≥2 Gy	<2 Gy	<2 Gy
Number of fractions/day	1	1	2–3	2
Treatment days/week	5	6	5	5–6
Overall treatment time	7 weeks	5 weeks	7 weeks	<5 weeks

The 10-year, follow-up report of a phase III randomised trial (SAKK 10/94) treating advanced HNC patients with concomitant cisplatin and hyperfractionated radiotherapy versus hyperfractionated radiotherapy alone showed that all endpoints in the combined regimen were superior, without differences in late toxicities (Ghadjar et al. 2012). Therefore, locoregional failure-free survival, distant metastases-free survival and cancer-specific survival were all improved in the chemoradiotherapy arm, where patients were treated with 74.4 Gy, 1.2 Gy twice daily 5 days a week, combined with two cycles of cisplatin 100 mg/m^2 over 5 days in weeks 1 and 5. Similar outcome was reported by Budach et al. (2015) after evaluating the results of the ARO 95-06 randomised phase III trial, where patients were randomised to be treated with either hyperfractionated accelerated radiotherapy of 70.6 Gy with concurrent chemotherapy (5FU/mitomycin C) or with radiotherapy alone (77.6 Gy hyperfractionated accelerated radiotherapy). At 10 years, locoregional control rates in the combined treatment arm were 38% versus 26% in the radiotherapy arm (p = 0.002), while cancer-specific survival rates were 39% versus 30% (p = 0.042).

5.5 CANCER STEM CELL-TARGETING MECHANISMS

Cancer stem cells were shown to exhibit higher resistance to treatment than non-stem cancer cells (Moore & Lyle 2011). This property on its own would justify specific targeting of CSCs in order to achieve tumour control. However, besides their intrinsic resistance, another reason for their endurance is the CSC environment, given that these cells preferentially reside in specific microenvironmental niches within the tumour (Brunner et al. 2012). For instance, head and neck CSCs were shown to exist in perivascular niches (Krishnamurthy et al. 2010). Therefore, CSCs should not be the only clinical focus when designing a new treatment approach, but the stem cell niche must be targeted as well.

To date, several CSC-targeting mechanisms and pathways have been identified (Figure 5.5) and it is expected that others will be revealed in the near future.

Some of the targeting pathways that are currently under investigation are presented in the following sections.

5.5.1 Signalling Pathways

Molecular signalling pathways in normal stem cells play a crucial role in cell proliferation, differentiation, survival and self-renewal, whereas their dysregulation can lead to carcinogenesis. The most important signalling pathways that were identified to be linked to the role of CSCs in tumorigenesis are Notch, Hedgehog (Hh), and Wnt pathways (Muller et al. 2007).

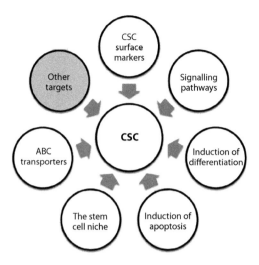

FIGURE 5.5 Cancer stem cell targeting mechanisms.

Notch signalling is mainly involved in cellular communication, including CSCs, immune cells and vascular endothelial cells, and has several functional roles. Crosstalk with other pathways can also dictate the overall effect of Notch signalling. While Notch was found to be oncogenic in several tumour types, HNC mutations identified in Notch receptors suggested a tumour-suppressive function (Agrawal et al. 2011). The role of Notch1 as a tumour suppressor in HNC was suggested due to the loss-of-function mutations detected in a subgroup of patients. A more recent study revealed a dual role of Notch1 in HNC (Rettig et al. 2015), which varies from tumour to tumour among HNC, based on the pattern of NICD1 (Notch1 intracellular domain) expression, which is also associated with clinical outcome. Thus, while in a subgroup of HNC patients the Notch pathway activation has tumour suppression ability, in another subgroup has oncogenic properties. From a clinical perspective, Notch-targeted therapies could be personalised based on biomarkers of Notch1 activity (Rettig et al. 2015).

The role of Hh signalling pathway is critical in stem cell development and maintenance. High expression of Sonic Hh (one of the three ligands of Hh signalling) was shown to significantly correlate with poor treatment outcome in HNC patients and could serve as a prognostic factor for these patients (Schneider et al. 2011). Overexpression of Hh is common in several malignancies, including HNC and is involved in epithelial mesenchymal transition, carcinogenesis and tumour spread (Gan et al. 2014). In several cancers, post-treatment transformation of epithelial cells into a mesenchymal phenotype that is known to increase resistance to therapy was associated to the Hh pathway. A retrospective evaluation of the RTOG 9003 trial showed that those HNC patients that expressed high levels of Gli1 before treatment had poor local control and overall survival as well as higher rates of distant metastases (Chung et al. 2011). Moreover, increased expressions of the Hh effector transcription factor Gli1 were shown to be further elevated by irradiation, leading to treatment resistance (Gan et al. 2014). To target this pathway, Hh inhibitors should be considered as a strategy to radiosensitise HNC cells.

Another signalling pathway essential for stem cell regulation and tumorigenesis is the Wnt/β-catenin signalling. There is experimental evidence showing that several regulatory components of the Wnt pathway are dysfunctional in HNC, which results in increased levels of nuclear β-catenin. Overexpression of β-catenin in HNC was shown to induce cell proliferation and dedifferentiation, i.e. transformation of non-stem cells into stem-like cancer cells (Lee et al. 2014). On the other hand, knockdown of β-catenin in HNC stem-like cells inhibited their self-renewal ability, treatment resistance and in vivo tumorigenicity. High levels of β-catenin in HNC patients are therefore correlated with poor prognosis.

Although the role of Wnt/β-catenin in HNC is not fully elucidated, more research into the inhibition of the aberrant Wnt signalling pathway would certainly have a positive impact on treatment outcome. Rudy et al. (2016) have investigated in vivo techniques to inhibit the Wnt pathway of human squamous cell carcinoma. By employing a small molecule inhibitor of the Wnt pathway, WNT974, CAM assays (chicken chorioallantoic membrane inoculated with cancer cells) were used to test the effectiveness of the agent. WNT974 was shown to disrupt the growth of several xenografts and also the development of distant metastases of several squamous carcinoma cell lines implanted in chicks (Rudy et al. 2016). Furthermore, an interaction between the Wnt and Notch signalling pathways was observed. It was suggested that tumours that exhibit Notch1 mutations, have increased Wnt signalling and are more prone to Wnt pathway inhibition. This is in agreement with earlier observations whereby HNC cell lines with loss-of-function mutations in Notch1 presented a higher response rate to WNT974 (Liu et al. 2013). Once again, this observation shows the complexity of signalling pathways and the influence of their interaction on treatment outcome.

5.5.2 The Stem Cell Niche

Similar to normal stem cells that reside in a stem cell niche to preserve their properties, CSCs live in their own microenvironment that provides the required signalling pathways to regulate common homeostatic processes and self-renewal. Niches are microenvironments that allow complex interactions between cells and matrix components in order to conserve stemness. HNC cells were shown to be mainly localised in close proximity to tumour blood vessels, i.e. in the perivascular niche, within a 100-μm radius of blood vessels (Krishnamurthy et al. 2010; Ritchie & Nör 2013). In vivo experiments demonstrated a strong correlation between CSCs and vascular endothelial cells, proven by the selective ablation of endothelial tumour cells which resulted in a significant decrease in CSCs (Krishnamurthy et al. 2010). This observation suggests that anti-angiogenic therapies could be effective on targeting CSCs and they should be part of therapeutic strategies for HNC.

As already mentioned in Section 5.3.1, CSC often reside in a quiescent state for self-preservation, where they are also protected against treatment. CSCs are refractory to traditional chemotherapy as most agents are targeting cycling cells by interfering with their division or DNA synthesis. This leads to tumour dormancy which in time can drive cancer relapse (Kusumbe & Bapat 2009). The identification of the mechanisms that sustain the dormant state and also the switch between the quiescent and proliferative states of CSCs could have important therapeutic implications.

5.5.3 Cancer Stem Cell Differentiating Agents

Since CSCs generally present with a higher resistance to treatment than differentiated cells, a possible strategy to increase tumour response to therapy would be to stimulate CSC differentiation. One agent that was found to display a powerful differentiating potential is all-trans-retinoic acid (ATRA), a member of the retinoid family and an active metabolite of vitamin A (Lim et al. 2012; Bertrand et al. 2014).

Retinoids exhibit a powerful effect on cell differentiation and apoptosis, by influencing multiple signalling pathways that are involved in stem cell preservation (Campos et al. 2010). Cellular differentiation caused by ATRA was associated with downregulation of the Wnt pathway, which is a key mechanism that controls malignant transformation. As supported by both in vitro and in vivo preclinical studies, the complexity of DNA damage caused by ATRA has induced cell cycle arrest and apoptosis in HNC (Lim et al. 2012; Bertrand et al. 2014). When combined with radiation, the effect of ATRA resulted in a decrease in the cell surviving fraction (Bertrand et al. 2014).

Retinoic acid was shown to be effective as a differentiating agent in nasopharyngeal carcinoma, by increasing IKKα expression, the enzyme involved in cellular response to inflammation (Yan et al. 2014). The group showed that reduced IKKα expression is responsible for the undifferentiated phenotype of nasopharyngeal cancer and its overexpression, stimulated by retinoic acid, induces differentiation and reduces tumorigenicity.

Another group of agents that are currently in focus are the histone deacetylase inhibitors (HDACi), that have the potential to inhibit tumour cell proliferation by inducing differentiation, cell cycle arrest and apoptosis (Chikamatsu et al. 2013), similar to retinoic acid. Alterations of CSC phenotype in HNC were achieved with HDACi by decreasing the mRNA expression of stemness-related genes and also by suppressing the epithelial mesenchymal transition phenotype of CSCs. Successful growth inhibition of HNC cells was attained with HDACi, which is indicative of therapeutic potential for CSCs.

5.5.4 Other CSC Targets

Apart from the above described targets, there are other potential therapeutic targets that include CSC markers such as surface markers, enzymes (ALDH) and drug efflux-pumps (ABC transporters) (Bütof, Dubrovska & Baumann 2013).

MicroRNAs are also important regulators of stem self-renewal and differentiation and they can act as both tumour suppressors and oncogenes (Hatfield & Ruohola-Baker 2008). Experiments undertaken on nine HNC cell lines looking at the expression of 261 miRNA genes revealed that 33 miRNAs are highly expressed whereas 22 showed low levels in all studied cell lines. Also, miR-21 and miR-205 were more notably expressed and were suggested to act as oncogenes in HNC (Tran et al. 2007).

Induction of CSC apoptosis is another mechanism of tumour control. Since apoptosis dictates the developmental balance within a tissue, dysregulation of this mechanism can lead to cancer initiation and CSC resistance. While activation of the apoptosis signalling pathway would be required for CSC eradication, the resistance mechanisms to apoptosis in CSCs are not fully understood.

Due to the vital role of CSCs in tumour growth and development, CSC-specific targeting agents are needed to increase tumour control and decrease the rate of recurrence in HNC patients. Several potential targets have been identified so far together with several complex mechanisms that dictate the fate of the cell. While experiments in this direction show promising results, more research is needed to develop efficient targeted therapies likely to be clinically implemented (Krause et al. 2011). One of the greatest challenges is to develop a highly CSC-specific and directly targeting agent that can optimise the effect of radiotherapy without increasing normal tissue toxicity.

CHAPTER 6

The Radiobiology and Radiotherapy of HPV-Associated Head and Neck Squamous Cell Carcinoma

6.1 HPV+ HNC AS A NEW TUMOUR TYPE

For many years, HNCs were attributed to alcohol and tobacco consumption, where the synergistic effect of the two agents resulted in an increased incidence of this cancer group. However, over the last few decades, the incidence of tobacco-related HNCs has been declining due to the reduction in tobacco use. Nevertheless, the number of overall HNC cases does not show significant decrease, and this is credited to the increasing incidences of HPV (Chaturvedi et al. 2013).

The induction of HPV-related HNC was shown to be related to the same virus that causes cervical and anogenital neoplasms (zur Hausen 2000). Recent studies have identified a high-risk subset of HPV (HPV16) that arises due to genetic alteration activated by the HPV oncoproteins E6 and E7. These oncoproteins have the ability to inactivate the tumour suppressor genes p53 and Rb and induce aberrant epithelial proliferation (Werness, Levine & Howley 1990; Strati, Pitot & Lambert 2006) (Figure 6.1).

The most common types of HNCs are oral and oropharyngeal squamous cell carcinomas (Jemal et al. 2011). Furthermore, these incidences are on the rise due to the preferential localisation of HPV-induced lesions within the oropharynx. Studies looking at the prevalence of molecular HPV markers in oropharyngeal versus non-oropharyngeal HNC showed that the probability of oropharyngeal cancer being HPV-related is fivefold higher than that of non-oropharyngeal cancer (Combes & Franceschi 2014).

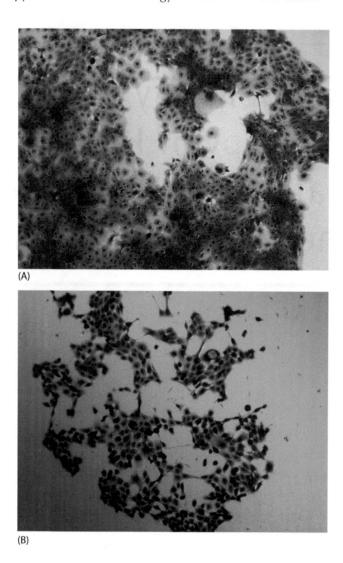

FIGURE 6.1   Images of HPV+ UM-SCC-47 cell line before (**A**) and after (**B**) irradiation with a 4 Gy 6MV photon beam. (Courtesy of Paul Reid & Eva Bezak, University of South Australia.)

As already mentioned in Section 2.5.7, the discovery in 1983 of HPV-associated HNC (Syrjänen et al. 1983) can be considered a landmark in head and neck oncology as this new entity differs in epidemiology, aetiology, treatment response, and prognosis from the HPV negative (HPV–) cancers.

The increased incidence of HPV positive (HPV+) HNC has introduced the need for a new approach in HNC management. The sections to follow tackle the radiobiological differences between the two HNC entities as well as the clinical implications of these disparities. Future treatment directions are also discussed based on the current scientific literature.

6.2 TESTING FOR HIGH-RISK HPV PATIENTS

Given the high and increasing incidence of HPV-caused head and neck squamous cell carcinoma (HNSCC) and considering the fact that HPV+ HNC patients show higher tumour control, and thus have a much better prognosis than HPV-patients, testing for high-risk HPV has become a prerequisite in all HNC trials. The importance of such testing is related to the early identification of this patient group and to the necessity for differential treatment in order to reduce unwanted side effects. Clinicians accept the concept that treatment should be administered according to the HPV status of each individual (mainly for oropharyngeal cancer patients), and trials are ongoing to identify the optimal therapy, whether combined or single agent.

Detection of transcriptionally active high-risk HPV is nevertheless, challenging, due to the lack of a consensus regarding the ideal diagnostic test (Table 6.1). As already mentioned in this chapter, HPV-related oropharyngeal cancers account for ~80% of all oropharyngeal neoplasms (Singhi & Westra 2010). On the other hand, <3% of HNC outside the oropharynx (i.e. oral cavity, larynx, hypopharynx) tested positive for transcriptionally active, high-risk HPV when using E6/E7 mRNA in situ hybridization (Lewis et al. 2012).

Consequently, HPV testing should mainly target the oropharynx patient group. Furthermore, since >90% of HPV+ oropharyngeal cancers are caused by HPV type 16, any testing approach must include this high-risk HPV type (Bishop et al. 2015). To justify routine clinical implementation and standardisation, a test/assay should fulfil the following objectives:

TABLE 6.1 Current HPV Testing Methods with Their Associated Advantages and Drawbacks

Method	Advantages	Disadvantages
p16 immunohistochemistry	Simple Widely available Cost-effective	Only recommended for oropharyngeal cancers (as stand-alone test) If p16 immunostaining is lower than the cut-off value (70% tumour cells), additional tests are required
Polymerase chain reaction	High sensitivity Widely used and considered the gold standard	Low specificity Technically challenging Time consuming Unable to identify the anatomical origins of the HPV infection
DNA in situ hybridization	High specificity Allows easy integration into laboratory Reliable detection and visualization of DNA	Limited sensitivity for samples with low viral copy numbers
RNA in situ hybridization	All the advantages of the DNA ISH Identifies transcriptionally active HPV High sensitivity	To further increase specificity, multimodality testing is required

- Accuracy (highly specific and sensitive)

- Minimal invasiveness

- Cost-effectiveness

- Time-efficiency

- Availability

Current HPV testing methods include p16 immunohistochemistry, polymerase chain reaction (PCR)-based methods, DNA/RNA in situ hybridization, or routine histology, which is often disregarded despite the fact that HPV+ HNSCC usually present with a specific morphological feature (Bishop et al. 2015).

One of the simplest, widely available, and most cost-effective HPV testing methods is *p16 immunohistochemistry*. This test detects the tumour suppressor protein, p16, which is a surrogate marker for transcriptionally active high-risk HPV infection. In HPV− HNSCC, p16 is usually inactivated by several genetic mechanisms including point mutation, homozygous deletion, and promoter methylation (Reed et al. 1996). Therefore, the presence of p16 correlates with HPV positivity. This test, as a stand-alone assay, is only recommended for oropharyngeal cancers, as for other HN-related anatomic sites, the specificity of p16 is low or unclear (Bishop et al. 2012). Even in oropharyngeal cancers, the test has its limitations, as for a positive interpretation, p16 immunostaining must be observed in at least 70% of tumour cells (Bishop et al. 2015). When lower percentages are identified, a supplementary test such as DNA PCR might be required.

Therefore, other highly sensitive and commonly employed HPV tests are the *PCR-based methods*. Within this technique, a signal sequence of DNA is detected and amplified several orders of magnitude, thus the method is extremely sensitive to the presence of HPV. Nevertheless, the specificity of the test for transcriptionally active HPV is low, as the simple detection of HPV DNA does not necessarily indicate that it has clinical significance, i.e., it is causative of the disease (Bishop et al. 2015). What is, however, expected from the PCR test is the mRNA identification of the oncogenic proteins E6 and E7 using quantitative reverse transcriptase PCR. Real time PCR amplification of viral E6 and E7 mRNA is considered the gold standard in detection of transcriptionally active HPV. Nevertheless, the method is technically challenging and time consuming, and it cannot identify the anatomical origins of the HPV infection (Venuti & Paolini 2012).

A molecular technique that allows reliable detection and visualization of DNA within the histological sample is *in situ hybridization* (ISH). In the DNA ISH technique, punctate nuclear signals can be identified in the tumour population which, evaluated microscopically, are indicative of the HPV presence. The test has a high specificity and can be easily integrated into the pathology laboratory. For tumour samples with low viral copy numbers, the test has limited sensitivity. Recent advances in technology led to the development of a sister test, the RNA ISH, which has the additional advantage of identifying the transcriptionally active HPV. Both the specificity and the sensitivity of the RNA ISH technique are high and it was shown to better correlate with patient outcome than DNA ISH (Ukpo et al.

2011). The RNA ISH method might be the test choice in the near future due to its several advantages over more established HPV tests.

Often, for a higher specificity, a multimodality testing approach is employed, the most common combination consisting of PCR for HPV DNA or in situ hybridization complemented by p16 immunohistochemistry.

One of the latest developments in the area of HPV testing is the *saliva-based DNA test* in patients with squamous cell carcinomas of the head and neck, which was shown to be specific for p16-positive oropharyngeal tumours (Chai et al. 2016; Wasserman et al. 2017).

A pilot study undertaken on 42 HNSCC patients aimed to compare the detection in salivary oral rinses of HPV16 biomarkers with tumour p16 expression using quantitative/reverse transcription PCR (Chai et al. 2016). The high specificity of this assay (100%) has demonstrated that the detection of HPV16 DNA in salivary oral rinse correlates with HPV status and can be employed as a non-invasive and cost-effective diagnostic tool in all high-risk individuals.

The above pilot study results have been confirmed by a recent study reported by Wasserman et al. (2017) using nested PCR in determining the HPV positivity, and immunohistochemistry for evaluating tumour p16 status in 62 HNSCC patients. The test was shown to be highly reliable as a positive high-risk HPV saliva assay, being 100% specific (95% CI), and had a 100% positive predictive value for p16 positive oropharyngeal tumours. Irrespective of the tumour site, the saliva assay had a specificity of 94% (95% CI) for a p16 positive tumour. This test fulfils several conditions of an ideal test, by being simple, fast, and reliable when biopsies are not readily available (Wasserman et al. 2017).

The number of tests available for HPV identification offers various choices for laboratories and current clinical practices. However, it is critical that in patients, only clinically validated HPV assays be used.

To emphasise the clinical importance of HPV testing among HNC patients, several national organisations have released, or have under preparation, guidelines for routine HPV testing of oropharyngeal cancer patients (Min & Houck 1998; Westra 2014). It is time for more standardised and feasible testing approaches to be embraced by the clinical community.

6.3 RADIOBIOLOGICAL DIFFERENCES BETWEEN HPV– AND HPV+ HNC

Before discussing the clinical implications of differences between HNSSC based on the presence of HPV, the underlying radiobiological behaviour of HPV+ HNC is examined. The results of preclinical studies looking into the radiobiological features of HPV-infected cell lines lead us to the building blocks of radiotherapy, as defined by Withers (1975) and complemented by Steel, McMillan and Peacock (1989):

1. Repair of sublethal injury in normal and neoplastic cells

2. Reoxygenation of the tumour

3. Redistribution through the division cycle

4. Regeneration of surviving normal and malignant cells between dose fractions

5. Radiosensitivity

6.3.1 Repair of Sublethal Injury in Normal and Neoplastic Cells

Ionising radiation can induce single-strand breaks, which are usually repaired with high accuracy; and double-strand breaks, which are potentially lethal lesions. Sublethal lesions can possibly be repaired, unless a subsequent dose of radiation transforms the lesion into an unrepairable damage. Sublethal damage repair is a critical aspect concerning normal tissue recovery during treatment. While this process is crucial in diminishing normal tissue-related adverse events, repair of sublethal injury is not sought within the tumour population.

Nevertheless, repair commonly occurs among tumour cells between fractions of radiotherapy. A chromosomal instability syndrome that is associated with cancer predisposition is Nijmegen breakage syndrome (NBS). NBS1 is the NBS gene product that is responsible for the DNA double-strand break repair (Karran 2000). It was determined that overexpression of NBS1 increases cell proliferation and promotes tumorigenesis (Chiang et al. 2003). Overexpression of NBS1 is found in about half of advanced HNC patients. In HPV-cell lines, the increased expression of the DNA double-strand break repair gene NBS1 was found to be an independent marker of aggressive HNSCC and poor prognosis (Yang et al. 2006).

As compared to HPV-cell lines, irradiated HPV+ HNC cells were shown to present a large number of residual DNA double-strand breaks (Busch et al. 2013; Arenz et al. 2014). This observation implies that HPV+ cells exhibit a compromised repair ability, which contributes toward their greater radiosensitivity (Rieckmann et al. 2013). Experiments undertaken on HPV p16+ cell lines have shown that altered DNA repair ability is associated with a significant G_2 phase arrest/ which is linked to checkpoint protein kinase 1 (Chk1) activity (Busch et al. 2013). Further investigations revealed that Chk1 inhibitors can abrogate radiation-induced G_2 arrest and cause sensitisation to radiotherapy (Busch et al. 2017). The G_2 checkpoint Chk1 can, therefore, offer a therapeutic target in HPV+ head and neck tumours.

6.3.2 Reoxygenation of the Tumour

6.3.2.1 Hypoxia

Tumour hypoxia is a known factor responsible for treatment resistance and ultimately, failure. This aspect has been discussed in detail in Chapter 4. Tumour reoxygenation is, therefore, critical in order to achieve better treatment response. Fractionation of radiotherapy (and chemotherapy) facilitates reoxygenation in between subsequent doses of radiation (or drugs) due to tumour shrinkage and easier access to blood supplies. While this process encourages reoxygenation of chronically hypoxic areas, acute hypoxia often remains a challenge. The quantitative assessment of acute hypoxia in HNC relies on current and future developments of hypoxia-specific radioisotopes in positron emission tomography/computed tomography (PET/CT) and magnetic resonance (MR) imaging. Given the fact that hypoxia promotes angiogenesis and metastatic spread, the management of HNC must focus on techniques to efficiently identify and then to overcome tumour hypoxia.

Patient studies confirmed the finding whereby tumour hypoxia is independent of HPV status (Mortensen et al. 2012; Trinkaus et al. 2014). The relationship between hypoxia, HPV status and treatment outcome has been investigated within a clinical study using

FMISO-PET hypoxia-specific imaging (Trinkaus et al. 2014). The results showed high prevalence of hypoxia in both p16-positive and negative HNSCC (74% vs 80%), without differences in hypoxia distribution.

To investigate the link between hypoxia and the high radioresponsiveness of HPV+ HNSCC, experiments undertaken on cell lines indicated that while under normal oxygenation HPV+ cell lines are 2.4 times more radiosensitive than HPV– cultures, the two strains presented comparable radioresistance in hypoxic environment (Sørensen et al. 2013). Additionally, when the hypoxic cell radiosensitiser, nimorazole, was administered under hypoxic conditions, the sensitising effect of the drug (1.13–1.29) was independent of the HPV status. The conclusion drawn from this observation is that HPV+ patients might not benefit from hypoxic sensitisers. However, this result is not because of differences in hypoxia sensitivity or drug response, but due to the overall higher radiosensitivity of HPV+ cells.

Further investigations of the effect of radiation on hypoxia in HPV+ HNC in vivo models have found an explanation to the limited effect of hypoxic cell sensitisers on this HNC group (Sørensen et al. 2014). Tumours, both HPV+ and HPV– HNC xenografts, treated with 10 Gy or 20 Gy were evaluated at different time points post-irradiation. While in the HPV+ tumour the hypoxic fraction decreased over time, this effect was not found in the HPV– tumour. The diminished hypoxic fraction, post-irradiation, in HPV+ cancers might explain the limited benefit from hypoxic modifiers that have been observed in patients (Sørensen et al. 2014).

6.3.2.2 Angiogenesis

Next to hypoxia, angiogenesis is another critical factor in tumour progression that influences treatment response. There is experimental evidence for disparities in angiogenesis among HNCs, which is linked to their HPV status (Troy et al. 2013). Immunohistochemistry studies of the epidermal growth factor receptor (EGFR) in various HNSCC show that EGFR exhibits higher expression in HPV– than in HPV+ tissues (Kong et al. 2009). EGFR is correlated with the angiogenic process via the activation of STAT3 (signal transducers and activators of transcription) which in turn, induces transcription of the vascular endothelial growth factor (VEGF). VEGF is known to be overexpressed in HNSCC and is associated with poor outcome. Interestingly enough, Troy et al. (2013) have found no difference in VEGF expression between HPV+ and HPV– tumour samples, a result that is suggestive of the fact that in HPV+ HNC the angiogenic process is less reliant on EGFR.

Elaborating on the idea of angiogenic differences, current experimental studies have demonstrated a correlation between the tumour suppressor protein p16 and angiogenic factors in HNC (Baruah et al. 2015). Pro-angiogenic factors (angiopoietin-1) and VEGF were found to a significantly higher extent in p16-negative patients compared with the p16-positive group. Furthermore, the main source of VEGF in p16-positive tissues consisted of tumour stromal cells, while in p16-negative tissues, they originated from both stromal and tumour cells. As mentioned in Section 6.2, overexpression of p16 in HNSCC is associated with HPV+ cancers, therefore these findings add to the understanding of radiobiological differences between HPV+ and HPV– HNSCC that could be exploited in future clinical trials.

Nevertheless, the inconclusive results concerning the extent and difference in the VEGF in HPV+ and HPV– tumours require further investigation.

6.3.3 Redistribution through the Division Cycle

From a radiation perspective, redistribution of cells along the cell cycle dictates overall survival, given that cells along the mitotic cycle exhibit phase-specific radiosensitivity. From the chemotherapy point of view, redistribution has an important outcome, considering that several chemotherapeutic agents display phase specificity. Therefore, the pattern of cellular passage through the division cycle influences overall responsiveness to treatment.

Experimental studies undertaken on four HPV+ and four HPV– cell lines showed that HPV+ cells progress faster through the radioresistant S phase, and accumulate in a more radiosensitive phase G_2/M (Arenz et al. 2014). As an effect, the amount of radiation-induced, residual double-strand breaks is higher and cell death is more pronounced than in the HPV– population. Therefore, it is suggested that a reason for the higher radiosensitivity among HPV+ tumours is cell cycle dysregulation, a process that causes cells to approach partial synchronisation in the more radiosensitive phases during radiotherapy (Arenz et al. 2014).

6.3.4 Regeneration of Surviving Normal and Malignant Cells between Dose Fractions

Surviving cells need adequate time for sublethal damage repair and repopulation, processes that are attainable during fractionated radiotherapy. Cellular regeneration of the normal tissue is critical for reducing unwanted early side effects. Nevertheless, malignant cells also have the opportunity to regenerate and proliferate in between fractions, which often leads to treatment failure. This aspect of tumour proliferation has been discussed in details in Chapter 5.

An in vivo study that has been mentioned in Section 6.3.2. regarding the link between hypoxia and HPV status, has also looked into proliferation behaviour of HPV+ and HPV– tumours when treated with 10 Gy or 20 Gy doses of radiation (Sørensen et al. 2014). HPV+ tumours showed a significant post-irradiation decline in cell proliferation levels, whereas in HPV– tumours this effect was not observed. Given the fact that tumour repopulation is one of the main culprits for treatment failure, this experimental observation supports the superior clinical response of HPV+ HNC.

Over the last couple of decades, it has become clear that cancer stem cells (CSC) are the main driving force in tumour development and progression. In their efforts to assess the role of HPV in treatment response and to find the rationale behind the superior responsiveness of this tumour group, researchers have explored the link between CSC and HPV. In a study undertaken on HNC cell lines and inoculated mice, no differences between the proportion of CSC among HPV+ and HPV– tumours was observed (Tang et al. 2013). An immediate conclusion resulting from this observation would be that the CSCs are not a strong influencing factor in treatment response, which would be a completely inaccurate statement given the powerful properties of CSC in regenerating the tumour. However, a paradoxical result was obtained by another recent study that showed elevated intrinsic CSC population in HPV+ as compared to HPV– tumours (Zhang et al. 2014). This outcome suggests that the CSC phenotype must play a role in tumour development and response to treatment, therefore the phenotype rather than the fraction of CSC is the key player in radioresistance.

6.3.5 Radiosensitivity

6.3.5.1 Intrinsic Radiosensitivity

Fractionated radiotherapy aims to overcome several mechanisms of radioresistance: hypoxia, tumour repopulation and intrinsic radioresistance. An important signal transduction pathway that is upregulated by radiotherapy is the phosphatidylinositol-3-kinase (PI3-K)/protein kinase B (AKT) cascade. The PI3-K/AKT pathway was found to be involved in all main mechanisms of radioresistance in HNSCC (Bussink, van der Kogel & Kaanders 2008) which implies that targeting this pathway could possibly improve tumour radiosensitivity in aggressive HNC. Radiation-caused DNA double-strand breaks are mostly repaired via non-homologous endjoining (O'Driscoll & Jeggo 2006), a process that is mediated by the DNA-dependent protein kinase catalytic subunit (DNA-PKcs) (Lieber et al. 2003). The regulation of DNA-PKcs is dictated by EGFR signalling through the PI3-K/AKT pathway (Toulany et al. 2006), therefore the activation of EGFR enhances DNA repair by the non-homologous endjoining system. Inhibition of this pathway results in an increase of residual DNA double-strand breaks, thus in cellular sensitisation to ionising radiation (Toulany et al. 2005).

HNSCC commonly present with high expressions of EGFR, which correlates with poor outcome. Despite the identification of this target, inhibition of EGFR does not always lead to better treatment outcome, which shows that HNC is a highly heterogeneous disease that requires more attentive patient selection for targeted therapies.

One of the largest studies investigating the association between EGFR in HNSCC, HPV status and clinical outcome was reported by Young et al. (2011). The group has confirmed that EGFR positivity (using fluorescence in situ hybridization assay) is inversely correlated with treatment outcome in oropharyngeal cancers and HPV16 positivity is a strong predictor of positive outcome. Furthermore, EGFR positivity and HPV16 positivity were found to be mutually exclusive in oropharyngeal cancer (Young et al. 2011).

6.3.5.2 Experimental Determination of Radiosensitivity by Clonogenic Assays

Several cell line studies underpin the observation whereby HPV+ HNC are more responsive to radiation than their HPV– counterparts. Based on a comparative test on four HPV+ and four HPV– cell lines, Arentz et al. (2014) have shown that after 2 Gy radiation, the mean SF_2 value in HPV+ cells was 0.198, whereas in HPV– cells was 0.34 (p = 0.010).

When irradiated with 3 Gy, Rieckmann et al. (2013) have observed similar discrepancies among head and neck squamous cell carcinoma lines: while the surviving fraction among the HPV– cells was $SF_3 = 0.45$, the fraction of HPV+ cells after a dose of 3 Gy was $SF_3 = 0.28$.

6.3.5.3 Radiosensitivity and Smoking

Based on clinical reports, the epidemiology of HNC has drastically changed over the past two decades. Studies show that the decline in tobacco use had a positive impact on

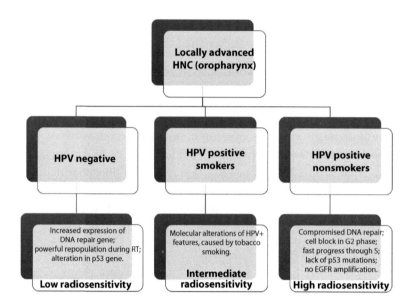

FIGURE 6.2   The multifactorial-dependent radiosensitivity of oropharyngeal squamous cell carcinoma as a function of HPV status and smoking history. (Adapted from Marcu. *Crit Rev Oncol/Hematol* 2016;103:27–36.)

tobacco-related HNSCC, while changes in sexual behaviour have increased the incidence of HPV+ HNSCC. As shown in the above sections, this latter etiological type of HNC has an increased radiosensitivity, thus a much better treatment response. Nevertheless, there is a subgroup of HPV+ HNSCC that present with a smoking history and an intermediate radiosensitivity given by the HPV-specific cellular changes that increase radiosensitivity but also due to tobacco-related molecular alterations that, on the other hand, lead to a decrease in radiosensitivity. Figure 6.2 presents the three HNSCC groups based on radiobiological differences and response to radiation.

The intermediate radiosensitivity of this patient group requires a different treatment approach from HPV+ HNSCC and should be administered the 'classical' HNSCC treatment. These aspects are discussed in section 6.4.

6.3.6 Radiobiological Effects of Combined Radio-Chemotherapy on HPV+ HNSCC

The above presented radiobiological differences between HNC as a function of their HPV status have exposed some of the underlying mechanisms explaining the higher radiosensitivity of those cell lines that tested positive for HPV. Nevertheless, pre-clinical studies indicate that similar effects are observed when cell lines are treated with combined chemo-radiotherapy. Of all chemotherapeutic agents used in HNC management, cisplatin continues to be the most commonly administered drug, thus the most frequently tested chemo-agent on cell lines.

A comparative study on HPV+ and HPV– cell lines revealed a synergistic effect of cisplatin when administered with radiation. Cisplatin was shown to greatly enhance apoptosis in HPV+ cell lines, suggesting that these entities are not only more radiosensitive but also more chemosensitive than HPV− cells (Ziemann et al. 2015). The underlying mechanisms, while not fully elucidated, are thought to be linked to (1) cell cycle dysregulation induced by cisplatin through S phase arrest, (2) cisplatin-induced downregulation of HPV E6 and E7 proteins that facilitates apoptosis and (3) increased therapeutic response of HPV+ cells to combined treatment, mediated by cisplatin (Ziemann et al. 2015).

Another, recent article has found a less significant difference of cisplatin's action on HPV+ as compared to HPV− HNC cell lines (Busch et al. 2016). The experiment showed that while the inhibition of proliferation and cellular toxicity was increased in HPV+ strains, this trend was not statistically significant when looking at the colony forming ability after incubation with cisplatin (p = 0.165) (Busch et al. 2016). The variations in response among HPV+ as well as HPV− cell lines were highly significant, which reflects the high inter-patient variability in treatment response in the clinics. This observation suggests the need for treatment personalisation through predictive assays with specific biomarkers and/or image-guided therapy throughout the treatment course.

Given the fact that cisplatin-based chemoradiotherapy is currently the standard of care in HNC management, regardless of the HPV status, the above results warrant further studies for optimal dose establishment in order to decrease normal tissue toxicity in responsive patients.

6.4 CLINICAL IMPLICATIONS OF DIFFERENCES BETWEEN HPV− AND HPV+ HNSCC

One major challenge in the management of HNC is the limitation imposed by the adverse effects. Treatment-related side effects in HNC patients are often debilitating, therefore any new protocol should consider reducing side-effects without hindering tumour control.

Tumours of the head and neck are often difficult to treat due to their anatomical location and their vicinity to critical organs. It is, therefore, important to consider the radiobiological differences between early and late responding tissues in the head and neck area for an optimal outcome. While the skin and the oral mucosa are early responding tissues, salivary (parotid) glands have a more complex radiobiological behaviour, presenting both early and late effects over four phases from 0 to 240 days (Coppes et al. 2001) and a reason why xerostomia is an omnipresent side-effect among radiotherapy patients treated for HNC. Having high radiosensitivity, salivary glands would benefit from dose de-escalation regimens in case of HPV+ patients. A clear demarcation between normal tissue characteristics and tumour behaviour must be considered for optimal treatment outcome.

6.4.1 Treatment Approaches

Recent and ongoing trials for HPV+ oropharyngeal cancers aim to reduce the aggressiveness of radiochemotherapy by dose de-escalation for both radiation and drug. The role of altered fractionation schedules in the management of HNC has been already presented and discussed. The choice of fractionation is strongly dependent on the radiobiological properties of the malignant cells. Therefore, while accelerated radiotherapy was employed to overcome tumour repopulation, hyperfractionation was designed to spare late reactions more, so allowing dose escalation and better tumour control. The clinical advantage of hyperfractionation over conventional treatment (and even over accelerated radiotherapy) was confirmed in the meta-analysis reported by Bourhis et al. (2006) and by the Radiatin Therapy Oncology Group (RTOG) 9003 trial, which is the largest fractionation study completed to date (Beitler et al. 2014).

The question arises whether hyperfractionated radiotherapy, with an overall smaller dose, would be an adequate schedule for HPV+ HNC, considering the radiobiological properties described above. Furthermore, given that radiotherapy as a single agent is less effective than combined with chemotherapy, future HNC trials for HPV+ patients should probably incorporate in their scheme the drug factor (Blanchard et al. 2011). Based on previous clinical results, several studies are investigating dose reduction practices for HPV+ HNC without compromising tumour response. The current reports advocate a dose range between 50 and 70 Gy with standard fractionation (i.e. 2 Gy fractions) correlated with tumour stage, smoking history, radiotherapy technique, combined treatment modality, and response to neoadjuvant chemotherapy (Urban, Corry & Rischin 2014).

Surgery remains an obvious choice for resectable HPV+ HNSCC. Lately, transoral robotic surgery is recommended instead of open surgery, due to better patient management and increased overall survival (Ford et al. 2014; Hammoudi et al. 2015). Nevertheless, the advantage of surgery over other treatment methods has not been demonstrated in HPV+ patients (Ang & Sturgis 2012).

Several studies have investigated dose de-escalation methods in various treatment combinations. While the number of agents tested in HNC to increase tumour control is large, only a few have proven their worth in HPV+ cancers. For instance, while cetuximab needs more clinical attention as an anti-EGFR agent (see currently ongoing randomised trials NCT01302834, NCT01874171 and TROG-1201-HPV), hypoxic cell sensitisers did not show any particular advantage in HPV+ cancer patients (Lassen et al. 2010; Rischin et al. 2010). The sections below discuss the role of neoadjuvant cisplatin and of cetuximab in the context of chemoradiotherapy.

6.4.1.1 The Role of Neoadjuvant Chemotherapy

The role of neoadjuvant chemotherapy (cisplatin) in HPV+ HNC is not clearly elucidated. This is partly due to the fact that there is only a limited number of studies that accrued oropharyngeal cancer patients—only to be treated with neoadjuvant cisplatin followed by concurrent (chemo)radiotherapy.

The French GETTEC (Groupe d'Etude des Tumeurs de la Tête Et du Cou) trial, has evaluated the effect of cisplatin-based neoadjuvant chemotherapy on overall survival in oropharyngeal cancer patients and have shown a considerably better outcome (5.1 years vs 3.3 years overall survival, p = 0.03) in the neoadjuvant chemotherapy arm as compared to the chemo-free arm (Domenge et al. 2000). The results of this study were so convincing that the trial was terminated before reaching the target sample size. Another study that focused on oropharyngeal cancer patients treated with neoadjuvant cisplatin followed by chemoradiotherapy has reported high tumour control rate (89%) which was though accompanied by severe toxicities (Finnegan et al. 2009). Nevertheless, the good overall outcome has justified further administration of this schedule in oropharyngeal cancer patients.

A retrospective study based on the TAX 324 trial has evaluated the treatment outcome for a subgroup of HPV+ oropharyngeal carcinomas treated with neoadjuvant chemotherapy followed by carboplatin-radiotherapy (Posner et al. 2011). This study compared two cisplatin-based induction chemotherapy regimens: docetaxel/cisplatin/fluorouracil versus cisplatin/fluorouracil, and found that the first drug cocktail was more effective. The 5-year overall survival reported among HPV+ patients was 82%, compared with the 35% in HPV− patients (Posner et al. 2011). Similar good results have been recently reported by a phase II UK trial of neoadjuvant chemotherapy and chemo-IMRT, showing that HPV+ HNC patients had significantly better outcome than HPV− patients, with 98.6% vs 76.8% locoregional progression free survival (Miah et al. 2015).

Majority of the above studies indicated that the maximum benefit from neoadjuvant chemotherapy was achieved by the subgroup of HPV16 oropharyngeal cancer patients.

A currently accruing phase III randomised clinical trial might have the decisional answer concerning the effectiveness of neoadjuvant chemotherapy on HPV+ oropharyngeal cancers. The Quarterback US study aims to compare two doses of definitive radiation therapy given with induction and concurrent chemotherapy in HPV+ oropharynx and nasopharynx (NCT01706939). The main goal is to evaluate the feasibility and clinical outcome of a reduced dose carboplatin-radiotherapy (56 Gy) compared with the conventional 70 Gy dose combined with carboplatin, following neoadjuvant chemotherapy.

In the past, trials have enrolled HNC patients irrespective of their HPV status and in retrospect is highly difficult, if not impossible, to assess treatment outcome with deficient data on site-specific HPV positivity. Therefore, if HPV16 oropharyngeal cancer patients indeed show superior treatment outcome after neoadjuvant chemotherapy, further patient stratification is required when treating HNC patients in order to avoid unjustified toxicities.

6.4.1.2 Chemoradiotherapy and Cetuximab

There is no doubt that cisplatin has been the 'queen' of chemotherapy in HNSCC. This title is still valid, despite several attempts to replace it with other platinum compounds, and various drug cocktails, to limit cisplatin's normal tissue toxicity. Cetuximab exhibits different side-effects from cisplatin, and combined with radiation results in less aggravated side-effects. However, the role of cetuximab in HPV+ cancer patients is yet to be determined. The results of the recent RTOG 0522 trial reported by Ang et al. (2014) have shown that cetuximab did not improve patient outcome when administered with concurrent accelerated cisplatin-radiotherapy. Furthermore, when looking at the HPV status, the trial showed no correlation between the differences in EGFR expression and outcome.

Currently, three randomised clinical trials aim to elucidate the effectiveness of cetuximab compared with cisplatin, in combination with radiotherapy and, at the same time, to evaluate the treatment related side effects in the two study arms (RTOG 1016 [US], De-ESCALaTE HPV [UK] and TROG 12.01 [AUS]). The ultimate aim of the trials is to determine the optimal treatment schedule for HPV+ oropharyngeal cancer patients. The radiotherapy schedule varies among the trials as follows: the RTOG employs accelerated intensity-modulated radiation therapy over 6 weeks (NCT01302834), the UK trial is planning to administer conventionally fractionated IMRT (NCT01874171) and the

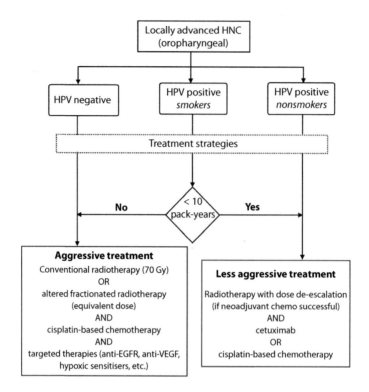

FIGURE 6.3 Chart representing current treatment approaches and projected protocols based on prognostic factors for oropharyngeal HNC. (Adapted from Marcu. *Crit Rev Oncol/Hematol* 2016;103:27–36.)

Australian trial is designed to use conventionally fractionated radiotherapy over 7 weeks (TROG-1201-HPV).

Given the fact that the UK and the Australian trials are still accruing during 2017, the final results will be expected in years to come. Although time-consuming, only well-designed randomised clinical trials will be able to guide the treatment of HPV+ HNC patients towards optimal results.

6.4.2 Patient Stratification

6.4.2.1 Smoking History and Risk Stratification

While HPV+ oropharyngeal cancers have been shown to represent an independent bio-clinical entity given by the particular epidemiology, biology and response to treatment, smoking was shown to have a strong influence on treatment outcome (Ang & Sturgis 2012; Broglie et al. 2013). Therefore, further patient stratification is required that would take into account, beside the HPV status, also the smoking history (see Figure 6.3).

Patients with HPV+ HNSCC with a history of >10 pack-years* tobacco smoking had poorer outcome than non-smoking HPV+ HNC patients (Broglie et al. 2013). These

---

* 1 pack-year is equivalent to 20 cigarettes per day during a year.

patients should therefore undergo similar treatment to their non-HPV counterparts (see also Figure 6.3).

When evaluating the smoking habit of HNC patients, the RTOG 0129 trial showed that the median pack-years of tobacco smoking among HPV+ smokers is considerably less than the median pack-years among HPV– patients (12.2 vs 36.5, p < 0.001) (Ang et al. 2010). Smoking was nevertheless found to be a significant determinant of progression-free and of overall survival. The risk of cancer recurrence or death increased by 1% for each additional pack-year of tobacco smoking in both HPV– and HPV+ cancer patients (Ang et al. 2010).

6.4.2.2 Tumour-Infiltrating Lymphocytes and Risk Stratification

Tumour-infiltrating lymphocytes (TIL) are those white blood cells that have left the blood stream and migrated into the tumour. TIL levels in the primary tumour were shown to trigger an immune response that has a powerful effect on treatment outcome (Ward et al. 2014). This observation confirms the fact that as a disease, HNC has a strong immunologic component. While the prognostic value of TIL is still debated in the literature, there are several studies that showed a positive link between the levels of TIL and treatment response in HNSCC.

The role of TIL in HPV+ patients has been investigated in a retrospective study on 270 oropharyngeal cancer patients (Ward et al. 2014). The study results showed that high levels of TILs conferred higher 3-year survival rate among HPV+ patients (96% for high TIL levels vs 59% for low TIL levels). It was determined that HPV+ patients that presented with low levels of TILs had similar survival rates as HPV– patients (Ward et al. 2014).

The predictive power of TIL in pretreatment specimens from 101 HNSCC patients, treated with chemoradiotherapy, was investigated within a larger study than previously published ones, thereby increasing its statistical significance (Balermpas et al. 2014). TIL immunostaining with several markers was scrutinised (CD3+, CD8+, CD4+ and FOXP3+ protein expression) and correlated with treatment outcome. While CD4 and FOXP3 expressions did not show any clinical relevance for patient stratification, elevated CD3 and CD8 expressions were associated with increased overall survival (Balermpas et al. 2014). The results of this study were confirmed by a more recent one, showing that high expression of CD8+ is a good prognostic factor for clinical outcome in both HPV+ and HPV– patients undergoing chemoradiotherapy, thus this TIL marker could be employed for treatment stratification (Balermpas et al. 2016).

In order to use TIL as a reliable factor for further HPV+ patient stratification, there is a need for standardization and establishment of a cutoff value for TIL levels. A recent study reported by Xu et al. (2017) aimed to ascertain a criterion to assess TIL levels and to investigate the association between TIL and prognosis. The study showed several correlations between TIL and clinical/patient-related factors: (1) TIL levels were directly associated with T stage and prognosis; (2) TIL was identified as an independent predictor of disease-free survival; (3) TIL levels showed inverse correlation with smoking and drinking habits; (4) Microscopic extracapsular spread was specific for patients with low TIL levels. The best threshold value for TIL levels that showed all the above correlations was found to be 70% (Xu et al. 2017). While more research is needed to confirm all the above results, TIL may be a valuable predictive factor for treatment outcome in HNSCC in the near future.

6.5 FUTURE MANAGEMENT OF HPV+ HNSCC

6.5.1 Conclusions Derived from Clinical (Radio)Biology Regarding the Responsiveness of HPV+ Oropharyngeal Squamous Cell Carcinoma

A number of conclusions can be derived concerning the differences in HNC behaviour as a function of HPV status, based on current pre-clinical and clinical evidence. The main mechanisms specific to HPV+ oropharyngeal cancers that are responsible for the superior treatment response as compared to HPV– cancers are presented in Table 6.2.

6.5.2 New Staging System for HPV-Related Oropharyngeal Squamous Cell Carcinoma

Because of the improved survival of HPV+ oropharyngeal cancer patients as compared to HPV– HNC, trials are ongoing to establish new treatment strategies that would reduce normal tissue toxicity without diminishing tumour control. An important component of a successful patient management is accurate tumour staging. The latest edition of the staging system recommended by the American Joint Committee on Cancer, the *AJCC Cancer Staging Manual: Seventh Edition* does not consider the prognostic advantage of HPV+ oropharyngeal cancers. In order to overcome the limitations of the AJCC recommendations, Malm et al. (2017) have validated and compared two recently proposed staging systems for this patient group: The International Collaboration on Oropharyngeal Cancer Network for Staging (ICON-S) (O'Sullivan et al. 2016) and the University of Texas MD Anderson Cancer Centre (MDACC) staging systems. The ICON-S staging was shown to provide the most adequate stratification of overall and progression-free survival. This study which was based on 1907 patients with HPV+ oropharyngeal cancer from seven cancer centres across Europe and North America, will highly likely serve as future guideline for staging.

6.5.3 Treatment Directions for HPV+ HNSCC

Although locoregional control and long-term survival rates among HNC patients have improved over the years, there is still plenty of room for further improvement. The role of medical imaging in diagnosis, treatment planning and guidance is an acknowledged fact. The prognostic markers that are most commonly employed for HNC involve the staging system that has been perfected due to the modern imaging techniques and the evolving pool of target-specific markers. PET/CT is a clear example here, where hypoxia-specific and proliferation-specific radiotracers are continuously investigated and trialled.

Nevertheless, the staging system does not always offer an accurate clinical prognostic (Lydiatt, Shah & Hoffman 2001), a fact that forces the research community to look into other alternatives. Molecular markers could be a potential solution in this direction, as also shown in the current chapter. Therefore, molecular alterations due to oncogene and tumour suppressor gene activations that manifest via gene amplification and overexpression of oncogenes, could serve as predictive markers in HNSSC. The EGFR, p53, c-myc, cyclin D and ras, are a few of the commonly studied targets that could improve clinical prognosis, given that their expression and identification in tumours is precise. While such molecular markers might play a critical role in the future, to date, the amount of conflicting reports place their routine use on hold (Hardisson 2003).

TABLE 6.2   The Main Mechanisms Specific To HPV+ Oropharyngeal Cancers that are Responsible for Their Radio/Chemo-Responsiveness, Based on Experimental Evidence

Mechanism/Factor Responsible for Superior Treatment Outcome	The Rationale behind the Mechanism/Effect on Treatment Response
(Radio)Biological Mechanisms/Factors	
Compromised DNA damage repair ability	HPV+ HNC cells have a compromised double strand break repair capacity that results in increased radioresponsiveness (Rieckmann et al. 2013).
Increased cell arrest in G_2 and faster progression through S phase	Cell survival is dependent on the position of the cell in the cell cycle during treatment. Knowing that cells in the S phase are the most radioresistant, their fast progression through this phase limits the number of cells that are present in S during irradiation. The same principle applies for the G_2 arrest, only that cells in this phase are more radiosensitive, thus cell arrest in G_2 increases the overall radiosensitivity of the tumour population (Lassen et al. 2014).
Heterogeneity of cancer stem cells phenotype	The fraction of cancer stem cells in HPV+ tumours was found to be equal or even higher than in HPV– tumours. Knowing the powerful effect of CSCs on tumour progression, this paradoxical result implies that the CSC phenotype rather than its proportion within a tumour dictates treatment outcome (Zhang et al. 2014).
Decreased EGFR protein expression	Overexpression of EGFR is a predictor of poor clinical outcome. EGFR positivity and the expression of p16 were found to be mutually exclusive in oropharyngeal cancer. Increased EGFR gene copy number is mainly limited to p16-negative cancers (Young et al. 2011).
Decreased amount of pro-angiogenic factors	Pro-angiogenic factors were found to a lesser extent in HPV+ tissues, which correlates to a better overall outcome (Baruah et al. 2015).
Decreased hypoxic fraction and cellular proliferation after irradiation	In vivo HNSCC studies showed a reduction in both cellular proliferation and hypoxic fraction after irradiation, which was particular for HPV+ tumours (Sørensen et al. 2014).
Clinical and Patient-Related Factors	
Younger age/better immune system	HPV+ HNC patients are usually young, with a better immune system and less comorbidities than HPV– patients. Experimental studies showed that HPV+ cancer patients are more immunogenic (Ang et al. 2010).
Non-smoking habits	HPV+ HNC patients are typically non-smokers. Smoking was associated to a poor treatment outcome due to a more aggressive tumour (Ang et al. 2010).
Increased levels of tumour-infiltrating lymphocytes	A number of studies showed that HPV16 patients have, on average, elevated levels of T-lymphocytes that correlate with better treatment responsiveness. More studies are needed in this direction as to date, there is no clear demarcation between tumour-infiltrating lymphocyte (TIL) levels in HPV+ and HPV– patients, thus TIL might be independent of HPV status (Ward et al. 2014).
Smaller primary tumour size	The reasons for smaller primaries among HPV+ HNSCC are justified by better tumour localisation and the lack of field cancerisation (Friedman et al. 2014).

CHAPTER 7

Normal Tissue Tolerance

7.1 THE RADIOBIOLOGY OF NORMAL TISSUES: EARLY AND LATE EFFECTS

The toxicity of HNC treatment has long been a concern in the management of this disease. Successful schedules in radiotherapy were developed empirically over the years with a cautious approach balancing increases in tumour control against toxicity levels that were considered acceptable. The later use of modified schedules or combined treatment modalities has allowed new advances in tumour control, but at the price of increased toxicity. Therefore, maintaining a delicate balance between tumour control and normal tissue toxicity requires good knowledge on characterising the magnitude and the impact of toxicity. Accurate toxicity assessment and analysis has been considered a critical area of oncology and practitioners are encouraged to familiarise with toxicity scoring systems and reporting standards to improve the quality of existing data (Trotti 2000).

Toxicity is a broad term describing the temporary and permanent changes in the appearance and functioning of normal tissues, including the related symptoms, which are caused by the treatment. It is important to note that no consensus terminology exists and alternate terms, such as effects, reactions, complications, morbidity, etc., are often encountered in the literature.

Two types of effects could be identified in normal tissues with respect to the occurrence in relation to the treatment duration. Thus, one could distinguish between acute effects that appear during treatment or within 90 days since the commencement of radiation therapy and late effects that appear after this interval (Cox, Stetz & Pajak 1995). However, the threshold is arbitrary and relates mainly to conventional fractionation schedules delivered over 40–50 days, for which epithelial effects appearing during treatment are usually healed 20–40 days after treatment. It is important to note that the introduction of aggressive fractionation schedules and the addition of chemotherapy have led to prolonged acute effects extending beyond the 90-day interval traditionally used. Nevertheless, the division of the normal tissue effects into acute and late could also be related to the proliferation features of the tissues involved. Thus, early reactions appear in rapidly proliferating tissues, while late reacting reactions appear in slowly proliferating tissues.

Although acute reactions are better characterised and understood than late reactions, it is thought that the mechanisms leading to their appearance are similar. Thus, radiation induces damage to the cells in normal tissues, which, when expressed, leads to cell depletion in the affected tissues that, in turn, determines the loss of tissue function. The time to express the damage depends on the proliferation features of the tissue and; therefore, rapidly proliferating tissues have an early manifestation of the effects, while slowly proliferating tissues have a delayed manifestation of the reactions. This stresses the importance of prevention of late effects, since the delay in manifestation means that only symptomatic treatments can be administered after the completion of the treatment.

The loss of function also depends on the organisation of the tissues and this leads to different tolerance levels for various tissues. Indeed, many tissues are organised in functional subunits (FSU) that perform part of the tissue function (Withers, Taylor & Maciejewski 1988). Knowledge on the FSU was obtained from emergency medicine and surgery when it was observed that tissues could continue functioning, even when parts were being removed. The FSU could be defined functionally as the unit of cells capable of performing the specified function of the tissue and in some cases, they could even be identified structurally as is the case of nephrons in the kidney. The organisation of the FSU in tissues could be parallel when each subunit is independent from other subunits, serial when the subunits are in strong dependence or a combination of these. Tissues with parallel organisation could withstand a large amount of damage due to their large reserve capacity, in contrast to serially organised tissues where damage to one FSU is enough to impair tissue function. While some organs have been identified as having a large reserve capacity similar to parallel tissues, as is the case with the lungs or liver; and others have more of a linear structure reminiscent of serial tissues, as is the case with the spinal cord or the oesophagus and other tubular structures of the digestive tract; it is important to point out that no tissue has been shown to have a purely parallel or serial organisation. Models have been proposed to describe the probability of inducing complications in normal tissues to account for this mixed organisation of tissues (Källman, Ågren & Brahme 1992). Another concept that is used to describe the responses of tissues to irradiation is that of tissue rescuing units (TRU) representing the minimum number of FSU, which is capable of rescuing the whole tissue from failure or from a specified level of injury. That has also been used in normal tissue complication probability (NTCP) modelling (Hendry & Thames 1986).

7.2 ACUTE REACTIONS IN HNC TREATMENT

Most acute reactions encountered from the radiation therapy of HNCs appear in the epithelial tissues. Damage in the basal keratinocytes of the skin results in erythema, while cell depletion of the epithelial basal layer of internal mucosae leads to mucositis. One could distinguish between mucositis, which is defined as a reaction in the treatment fields and stomatitis, which is chemotherapy-related toxicity appearing outside of the radiation fields for combined treatments.

In the case of the skin, cell depletion in the basal layer leads to a thinning of the epidermis, triggering an inflammatory response characterised by a secretion of cytokines, chemokines, histamine, and serotonin (Ryan 2012). Many of these substances have vasoactive properties determining an increased blood volume due to capillary dilatation leading to acute erythema of the skin (McQuestion 2011; Nystrom et al. 2004). Earlier symptoms also include discomfort, itching, and pain. More severe damage to the epidermis could lead to desquamation and even necrosis. Damaging the skin barrier also increases the risk for bacterial infections, further worsening the quality of life of radiotherapy patients.

Mucositis has a similar aetiology and appears pathologically as a small ulcer due to the depletion of the basal layer of the mucosa. This is usually covered by a pseudomembrane formed during the wound healing response. The membrane is usually white or opalescent when hydrated, but could appear yellow or greenish in the event of an infection. Minor trauma could dislodge the membrane, which results in bleeding. If the ulcer enlarges, it could connect with adjacent ulcers, producing a confluent pseudomembrane in the irradiated mucosa. Stomatitis induced by chemotherapy has a similar appearance but rarely evolves beyond the patchy appearance (Trotti 2000).

Regeneration of the epithelia leads to the disappearance of the inflammatory reactions caused by radiation within a few weeks after the completion of the treatment. However, it should be mentioned that infections with yeast, bacteria, or viruses could delay the healing of the epithelia. Lesions produced by the profound depletion of the stem cell compartment of the epithelia require longer healing times and could even lead to scarring. Furthermore, deep ulcers could later evolve into necrosis, which is recorded as consequential late effects of the acute lesions (Dörr & Hendry 2001).

Clinical investigators also have to monitor the appearance of acute effects in a number of other tissues, such as the salivary glands (acute xerostomia), the pharynx and the oesophagus (dysphagia) or the larynx (voice change). Some of these effects could, in turn, trigger other effects that may worsen the quality of life of the patients. Thus, acute xerostomia and dysphagia could hamper the nutrient intake of the patients leading, in turn, to weight loss. Comprehensive catalogues of criteria for monitoring the appearance of toxicities are available from a number of sources together with proposals for standardisation of recording and reporting (Cox, Stetz & Pajak 1995; Trotti 2000; Trotti et al. 2003).

Recording and reporting of toxicities from treatments are not straightforward. The incidence of acute effects following cancer treatment could be extracted from the records of clinical observations and patient symptoms. These could be supplemented by results from laboratory analyses, as well as records for additional procedures to which the patients have been subjected. Nevertheless, records could be affected by a large degree of interobserver and even intraobserver variability as has been shown in some comparative studies (Denham et al. 1996). These have highlighted the need for standardised criteria for toxicity grading and recording. Although normal tissue effects have been recorded since the beginning of radiotherapy, the standardised criteria for toxicity recording have appeared rather late with the Radiation Therapy Oncology Group (RTOG) recommendations (Cox, Stetz & Pajak 1995). The subsequent introduction of new therapeutic agents in combined treatments and observations of new toxicities have led to the introduction of the comprehensive Common Toxicity Criteria dealing with hundreds of toxicities in many organ categories from all therapeutic modalities (Trotti 2000; Trotti et al. 2003). Nevertheless, it has to be mentioned that mucositis and erythema are the most commonly reported acute toxicities from head and neck radiotherapy.

Standardised criteria have been adopted by many clinical trial organisations to improve consistency in the reporting of toxicity and comparisons among clinical treatment regimens. It has to be stressed that some criteria still have a qualitative character with less clear thresholds and they might suffer from the subjective evaluation of the observer. On one side, this highlights the need for training on clinical recognition and toxicity terminology into clinical programs (Trotti 2000) and, on the other, the need for quantitative methods for scoring toxicity. It has been reported that the assessment of skin reactions could be influenced by the chromatic acuity of the observers, the variation in ambient lighting, as well as other factors (Chu et al. 1960). Several quantitative methods have been explored, but their full potential still needs to be validated in systematic studies (Turesson et al. 1996; Momm et al. 2005; Tesselaar et al. 2017).

Typically, toxicity is reported on a four-grade scale corresponding to mild (grade 1), moderate (grade 2), moderately severe (grade 3) and severe (grade 4) reactions. A fifth grade has been associated with death, but this is a seldom event and is usually reported separately from other toxicity. Furthermore, intermediate grades are also used sometimes for better discrimination of the reported effects. Table 7.1 illustrates the RTOG scale used for reporting erythema in the skin of radiotherapy patients.

It has to be mentioned that comparisons of toxicities between studies are sometimes hampered by inconsistencies in reporting the observed rates of reactions. Thus, it is not uncommon to cluster the reactions into mild-moderate (Grades 1 & 2) and severe (Grades 3 & 4) and report them accordingly. Furthermore, some studies report only the severe (Grades 3 & 4) reactions, whereas others report only moderate (Grades 2 & 3) reactions (Trotti 2000). This situation illustrates some of the challenges that still have to be overcome for the uniform reporting and comparison of acute reactions.

Another confounding factor for the analysis of toxicity rates is the monitoring frequency of the patients. The usual practice with respect to acute reactions is to record and report the highest grade of toxicity experienced by the patient at any point during the 90 days' observation window. Nevertheless, the patients are usually seen by a clinician weekly during radiotherapy, every 3–4 weeks during chemotherapy and even more seldom after the completion of the treatment. It is therefore possible that some patients reach the peak of their reactions in between two observation sessions, leading to an underreporting of toxicity rates. Extending this reasoning, it is quite possible that clinical studies employing a higher monitoring frequency will report higher acute toxicity than those with sparser

TABLE 7.1 The RTOG Scale for Acute Erythema of the Skin

Grade 0	Grade 1	Grade 2	Grade 3	Grade 4
No change	Follicular, faint or dull erythema, epilation, dry desquamation, decreased sweating	Tender or bright erythema, patchy moist desquamation, moderate oedema	Confluent, moist desquamation other than skin folds, pitting oedema	Ulceration, haemorrhage, necrosis

Source: Based on Cox JD et al., *Int J Radiat Oncol Biol Phys* 1995; 1995;31:1341–1346.

TABLE 7.2   Acute Toxicity Data From Conventional Fractionation

Tissue	Grade 1	Grade 2	Grade 3	Grade 4
Skin	50%	35%	5%	0%
Mucous membrane	20%	45%	20%	0%

Source: Data from Lee DJ et al., *Int J Radiat Oncol Biol Phys* 1995;32: 567–576.

observations. Yet another confounding factor originates in the use of the highest observed grade, while it has been suggested that the timing and the duration of toxicity should also be accounted for (Trotti 2000). Attempts have been made to account for these aspects (Kaanders et al. 1992; Kaanders, Pop & Marres 1995; Denham et al. 1996), but they have yet to be widely adopted in the clinical practice of reporting toxicities.

The rates of acute reactions observed in the treatment of HNCs depend on several factors including dose, treatment duration, and the use of adjuvant therapies. Table 7.2 shows the approximate acute toxicity rates from head and neck radiotherapy employing conventional fractionation to 66–74 Gy (Lee et al. 1995).

Accelerated radiotherapy, either with conventional fractionation or with hyperfractionated schedules, has been shown to lead to increased acute reactions, but within acceptable limits (Horiot et al. 1992, 1997; Skladowski et al. 2000). However, extreme acceleration aiming to deliver two fractions per day or in a continuous course with 7 days a week has resulted in mucosal toxicity that has been deemed unacceptable leading to the stopping and modification of some clinical studies (Maciejewski et al. 1996; Jackson et al. 1997). The subsequent introduction of short breaks in the treatment has been shown to improve mucosal recovery through compensatory repopulation. However, the duration of the breaks has to be balanced against the decrease in tumour response due to tumour cell repopulation. The reader is directed to the chapter on radiobiological modelling for isoeffect calculations for acute reactions, for which an α/β value of 10 Gy is recommended (Thames & Hendry 1987).

The addition of chemotherapy in combined treatment approaches has been proven beneficial in the treatment of HNCs when given concomitantly as it increased tumour effect and overall survival (Pignon et al. 2000). However, the use of concomitant chemotherapy has also been shown to increase the acute reactions (Trotti 2000; Zackrisson et al. 2003). The magnitude of the increase seems to correlate with the type of chemotherapy. Thus, single-agent chemotherapy appears to lead to a small increase in severe mucositis (Browman et al. 1994; Bachaud et al. 1996; Al-Sarraf et al. 1998). In contrast, multi-agent chemotherapy combining cisplatinum and 5-fluorouracil have led to high rates of severe mucositis that required the introduction of breaks during the treatment, although without an increase in consequential effects (Denham & Abbott 1991; Adelstein et al. 1997; Brizel et al. 1998).

The use of prophylactic treatment of the acute side effects has also been investigated, but generally on small patient populations (Zackrisson et al. 2003). Although some trends have been identified in these studies toward a reduction in the acute effects, the magnitude of the effect was generally too small to recommend the approaches for standard practice (Franzén et al. 1995; Symonds et al. 1996; Meredith et al. 1997; Okuno et al. 1997; Lievens et al. 1998).

7.3 LATE REACTIONS IN HNC TREATMENT

Late effects from radiation originate either from chronic changes induced in the early reacting tissues or from pathophysiological changes induced by radiation in the vasculature and connective tissues. Thus, radiation induces vasodilatation and increases permeability of the blood vessels, which allow the deposition of fibrin and collagen that eventually leads to the formation of fibrosis in the vessel walls and the perivascular spaces. These, in turn, narrow the vascular lumen and decrease the vascular perfusion in a process consistent with arteroschlerotic changes due to aging, which ultimately result in an impairment of the function of the tissue that they support. The connective tissue pathogenesis has been related to increased collagen biosynthesis and enhanced proliferation of the irradiated fibroblasts, which subsequently invade the connective tissue leading to fibrosis (Cooper et al. 1995). Given the multitude of tissues involved in head and neck radiotherapy, late effects have a larger variability in terms of manifestation. Thus, damage of the mucosa or the skin could lead to chronic ulceration, atrophy and telangiectasia. Damage to the connective tissues could lead to oedema, fibrosis, trismus, and even necrosis. Damage to the salivary glands could lead to abnormal salivary flow, which, in turn, leads to dental decay. Damage to the nervous tissue could lead to neuropathies and necrosis. Damage to the bone or cartilages could lead to osteonecrosis or chondronecrosis.

Given the complexity of mechanisms leading to the appearance of late effects, it is important to note that reporting only acute toxicity is not enough to characterise the long-term consequences of various treatment approaches. This is an important aspect since the late effects are considered dose-limiting in radiation therapy. For head and neck treatments, most late effects appear within 3 years of treatment. This indicates that a follow-up period of at least 5 years is needed to record the incidence of late effects following HNC treatment. Furthermore, retrospective analyses of institutional records are unlikely to fully capture toxicity rates unless proper recoding practices are in place to capture the complexity of late effects (Trotti 2000).

Similar to acute reactions, standardised criteria are in use for recording and reporting late toxicity. The scoring scale of the RTOG (Cox, Stetz & Pajak 1995) has been much used for late toxicity recording, although it has been criticised as being insensitive to subtle changes. Alternative approaches include the Dische system (Dische et al. 1989), which was considered a more sensitive discriminator and has been adopted for some trials, and the SOMA (subjective-objective management analytic) criteria (Pavy et al. 1995).

The scoring scales also use a four-grade scale corresponding to minor (Grade 1), moderate (Grade 2), severe (Grade 3), and irreversible (Grade 4) damage. Death directly related to radiation late effects is sometimes included as the fifth grade in the scales (Cooper et al. 1995). Table 7.3 shows the approximate toxicity rates from head and neck radiotherapy employing conventional fractionation to 66–74 Gy (Lee et al. 1995).

The factors that influence the rates of late reactions observed in the treatment of HNCs include dose, dose per fraction, interfraction interval, and the use of adjuvant therapies. It is important to note that in contrast to acute reactions, there is no overall treatment time factor for the late reactions. The reason is the absence of compensatory proliferation from

TABLE 7.3   Late Toxicity Data From Conventional Fractionation

Tissue	Grade 1	Grade 2	Grade 3	Grade 4
Skin	35%	5%	1%	0%
Mucous membrane	20%	15%	2%	2%
Salivary glands	35%	35%	5%	0%
Pharynx & oesophagus	15%	10%	5%	1%
Larynx	15%	5%	2%	0%
CNS & spinal cord	3%	2%	0%	1%

Source: Data from Lee DJ et al., *Int J Radiat Oncol Biol Phys* 1995;32: 567–576.

the mechanisms of appearance of the late reactions. However, late reactions are much more sensitive to the dose per fraction and the interfraction interval due to the large capacity of the tissues involved for repairing sublethal damage (Cox, Stetz & Pajak 1995). Furthermore, clinical evidence has shown that nervous tissues are more susceptible to damage by altered fractionation (Zackrisson et al. 2003). This translates into a reduction of the rate of late reactions in hyperfractionated schedules. The increased sensitivity to fractionation of the late reactions is reflected by the low α/β values that characterise them compared with acute reactions. Thus, it is advised that isoeffect calculations should be performed using an $\alpha/\beta = 2$ Gy for nervous tissues and $\alpha/\beta = 3$ Gy for the others (Thames & Hendry 1987).

In case of accelerated treatments, the duration of interfraction interval should be carefully chosen in order to allow for repair of sublethal damage in late reacting tissues. Thus, it has been observed that very accelerated schedules like the CHART (Saunders 2010) delivering 36 fractions of 1.5 Gy, three times a day, have led to higher than expected late reactions since some of the interfraction intervals were as short as 6 h and did not allow for complete repair of sublethal damage from the first fraction by the time the subsequent fraction was delivered.

The effect of additional chemotherapy in combined treatment approaches on the rates for late effects has been studied much less than acute effects. However, indications exist that late effects are more frequent with combination therapy than with radiotherapy alone (Zackrisson et al. 2003). Furthermore, some agents have been observed to lead to permanent injury to peripheral nerves, inner ear, lungs and kidneys (Trotti 2000) and these factors have to be taken into account when judging the compliance of normal tissue burden with existing constraints.

Given the delayed onset of late reactions, much emphasis is put on their prevention, as few post-treatment interventions have shown any value. Thus, the most effective approach is to keep the radiation burden of late reacting normal tissues below the known tolerance levels in order to maintain as much as possible of the reserve functional capacity of the irradiated tissues. Some prophylactic measures could be used to further minimise the impact. These include exercising to decrease the incidence of normal tissue fibrosis and trismus, administration of pharmacological agents to increase tissue perfusion, to decrease inflammatory reactions, or to stimulate salivary flow and maintain good dental hygiene. It is important to point out that irradiation of normal tissues also poses difficulties for post-treatment interventions like surgery and tooth extractions as it increases the risk of tissue necrosis (Cox, Stetz & Pajak 1995).

7.4 NORMAL TISSUE TOLERANCE DOSES IN HNC RADIOTHERAPY

Accumulating knowledge on the dose constraints describing normal tissue tolerance to irradiation is a very slow process as it has to incorporate many variables with respect to dose, fractionation, timing of radiotherapy and not in the least the irradiation technique. Thus, much of the early work on skin and subcutaneous tolerance (Coutard 1932; Ahlbom 1941) was obtained from orthovoltage radiotherapy. The subsequent introduction of megavoltage therapy has brought to attention the tolerance of deeper seated normal tissues. Also, the non-uniform dose distributions in the normal tissues have changed with the irradiation techniques, which have evolved around the years from open fields to conformal radiotherapy and intensity-modulated radiation therapy. Technological changes have been paralleled by dose escalation studies and changes in fractionation patterns. All these changes have brought variability in dose distributions from which dose response relationships and tolerance levels have been derived. Nevertheless, determination of tolerance levels is still an evolving field as better dose calculation algorithms are taken into use and the synergistic effects of combined therapies have to be accounted for.

The first systematic collection of tolerance levels in radiation therapy has been published by Emami and colleagues (Emami et al. 1991). It included tolerance criteria for 28 sites of normal tissue, and for full and partial tissue irradiation. The tolerance list has been updated with dose-volume-outcome data accumulated from conformal radiotherapy as reviewed by the Quantitative Analysis of Normal Tissue Effects in the Clinic (QUANTEC) committee (Marks et al. 2010). Focused on conformal radiotherapy delivered with conventional fractionation, the QUANTEC analysis provides guidelines on clinically relevant endpoints, volume definitions, dose and volume limits as well as parameters for biological models for 16 specific sites. Table 7.4 summarises the dose-volume tolerance data for the organs at risk, relevant to radiotherapy of the head and neck.

While these tolerance levels have been derived from the current knowledge on therapeutic irradiation of normal tissues, care is advised when extrapolating the constraints for other treatment approaches, including hypo- or hyperfractionated schedules, intensity-modulated irradiation or particle therapy approaches. Indeed, calculation of tolerance doses for acute reactions would require accounting for overall treatment time when modifying the treatment schedule. Heterogeneous irradiation resulting from intensity-modulated radiation therapy might spare critical compartments of TRU, which will maintain the function of tissue even when conventional constraints are not fulfilled, as is for the example of salivary glands (van Luijk et al. 2015). Furthermore, particle therapy would have to account not only for highly heterogeneous irradiation of the normal tissues but also for the uncertainties in the relative biological effectiveness (RBE) of the particles compared with photon radiation (Dasu & Toma-Dasu 2013; Jones 2016; Kase et al. 2008).

Last, but certainly not least, the clinical use of normal tissue tolerance levels requires a delicate balance with maintaining target coverage, depending on the purpose of the treatment. Thus, the extent of the limitation of the radiation burden to the normal tissues might

TABLE 7.4   Dose Volume Constraints Relevant for Head and Neck Radiotherapy with Conventional Fractionation

Organ	Irradiated Volume	Dose or Volume Parameters	Complication Rate	Complication
Spinal cord	Partial	$D_{max} = 50$ Gy	0.2%	Myelopathy
	Partial	$D_{max} = 60$ Gy	6%	
	Partial	$D_{max} = 69$ Gy	50%	
Cochlea	Whole	$D_{mean} = 45$ Gy	30%	Sensory neural hearing loss
Parotid	Unilateral whole gland	$D_{mean} = 20$ Gy	20%	Parotid salivary function
	Bilateral whole glands	$D_{mean} = 25$ Gy	20%	reduced to <25% of
	Bilateral whole glands	$D_{mean} = 39$ Gy	50%	pre-RT level
Temporomandibular joint	Whole	$D_{mean} = 60$ Gy	5%	Limitation of joint
	Whole	$D_{mean} = 72$ Gy	50%	function
Skin, mucosa and conjunctive tissue		$D_{mean} = 55$ Gy	5%	Ulceration and necrosis
		$D_{mean} = 70$ Gy	50%	
Pharynx	Whole	$D_{mean} = 50$ Gy	20%	Pharyngeal constriction
Larynx	Whole	$D_{max} = 66$ Gy	20%	Vocal dysfunction
	Whole	$D_{mean} = 50$ Gy	30%	Aspiration
	Whole	$D_{mean} = 44$ Gy	20%	Oedema
	Whole	$V_{50\,Gy} = 27\%$	20%	Oedema
	Whole	$D_{mean} = 70$ Gy	5%	Cartilage necrosis
	Whole	$D_{mean} = 80$ Gy	50%	
Brachial plexus	Whole	$D_{mean} = 60$ Gy	5%	Clinically apparent
	Whole	$D_{mean} = 75$ Gy	50%	nerve damage

Source: Data from Emami B et al., *Int J Radiat Oncol Biol Phys* 1991;21:109–122; Marks LB et al., Int *J Radiat Oncol Biol Phys* 2010;76:S10–S19.

be reduced if maintaining target coverage is required to achieve a meaningful rate of local control. In other cases, dose coverage might have to be sacrificed to reduce the burden of normal tissues and maintain the quality of life of the patients.

7.5 RADIOPROTECTORS

As stated before, one major shortcoming of the often-aggressive head and neck radiotherapy is normal tissue toxicity, with mucositis, xerostomia, myelopathy and pneumonitis being some of the most common dose-limiting side effects. While acute adverse events can be efficiently managed with suitable medical care, late toxicities are often irreversible and are thus in need of better organ sparing. Besides accurate treatment planning, sparing healthy tissue during radiotherapy can be achieved, to a certain degree, with the addition of radioprotective agents that would, most importantly, act selectively upon normal and tumour cells. This selective uptake is probably the most critical aspect regarding radioprotectors, as the response of tumour cells should not be influenced by the addition of such agent. Furthermore, no additional toxicity should be prompted by the radioprotective agent in normal tissues.

To protect the healthy cells against radiation damage, the free radicals that create unrepairable damage must be restored to their original, stable state. Sulfhydryl compounds, which are endogenous hydrogen donors, compete with the oxygen molecule for the free radical. Naturally occurring sulfhydryl compounds are activated in the cell during their

DNA synthesis, a reason why cells in their S phase are more resistant to the effect of radiation. Consequently, the first radioprotectors have been developed from sulfhydryl compounds in the form of cysteine and cysteamine (Maisin 1989), although they had a very short lifespan due to toxicities when used in clinical concentrations.

Several groups of radioprotectors have been designed and tested (see Figure 7.1), but few made it into the clinics (Citrin et al. 2010).

Amifostine, previously known as WR-2721, is the only radioprotector that is in clinical use. The drug has been initially developed by the U.S. Walter Reed Army Institute of Research in the 1950s with the aim to protect military personnel against the effect of ionising radiation in a classified nuclear warfare project (Patt et al. 1949). Following project declassification, the drug was further investigated as a cytoprotective agent against cisplatin and radiation (Santini 2001). Amifostine is a prodrug that becomes active inside the normal tissues after dephosphorylation by the enzyme alkaline phosphatase (Calabro-Jones et al. 1985). The lack of this enzyme and/or its activity in tumours due to the low pH environment, leads to a selective uptake between healthy and cancerous cells. Other reasons for the selective uptake are the differences in the membrane structure between normal and malignant cells, which slows penetration of the drug into tumours, and also the poor vascularisation of tumours compared to normal tissues (Kouvaris, Kouloulias & Vlahos 2007). Experimental evidence shows that tumour protection due to amifostine uptake, if any, is weak (Koukourakis 2003).

In vivo animal studies have confirmed the ability of amifostine to diminish some of normal tissue toxicities induced by radio- and chemotherapy (Thomas & Devi 1987; Capizzi, Scheffler & Scein 1993; Cassatt et al. 2002). Similarly, some trials have proven the effectiveness of the drug in reducing certain treatment-related side effects, although the long list of normal tissues with preclinically demonstrable reduction of toxicities, was reduced to a much shorter one in clinical settings (Lindegaard 2003).

To quantify the effectiveness of amifostine in normal tissue protection, the dose reduction factor (DRF) has been determined for several tissues/organs as an expression of the

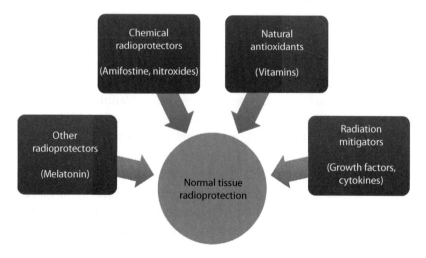

FIGURE 7.1 Different categories of normal tissue radioprotectors.

increase in radioresistance of the tissue under consideration (Hall 2000). Thus, DRF is calculated as the ratio between the dose of radiation in the presence of the drug and the dose of radiation in the absence of the drug, for the same biological effect. Some tissues of interest for head and neck radiotherapy are listed in Table 7.5.

As observed from Table 7.5, amifostine exerts a differential protection on various normal tissues. Nevertheless, the mechanism behind this preferential behaviour is not clear.

While in the past the results of clinical trials employing amifostine were inconclusive, and normal tissue protection was not always evident (Marcu 2009), a recent meta-analysis based on randomised controlled trials for HNC employing amifostine, could statistically confirm the clinical benefit of this radioprotective agent (Gu et al. 2014). The review encompassed 17 trials with 1167 patients, showing on a pooled data analysis that amifostine notably reduced Grade 3–4 mucositis, Grade 3–4 dysphagia, Grade 2–4 acute xerostomia as well as late xerostomia, without tumour protection and with manageable amifostine-related toxicities.

Nitroxide compounds have been investigated as possible radioprotectors over the last couple of decades, due to their ability to alter the tissue's redox status and to modify oxidative stress (Soule 2007). It was shown that both nitroxides radical and hydroxylamines (the reduction products), are recycling antioxidants that protect the cell against the oxidative stress, including the highly damaging hydrogen peroxide (Soule et al. 2007). The most representative nitroxide compound to date is *tempol* (4-hydroxy-2,2,6,6-tetramethylpiperidine-1-oxyl), which has been investigated in preclinical settings. In vivo animal studies showed that tempol was effectively protecting the salivary glands, without influencing tumour control (Cotrim et al. 2007), thus demonstrating a differential protection of normal tissues as compared to tumour cells.

Vitamin supplements, such as vitamin A, C and E, also administered as beta carotene, or alpha tocopherol, with strong antioxidant properties have also been investigated for their potential radioprotective effect on normal tissues. While their effect was generally lower compared with synthetic radioprotectors, the major concern about vitamins as radioprotectors is the poor differentiation between normal and tumour tissue protection, due to nonselective free radical scavenging (Citrin et al. 2010). This challenge has been confirmed by the results of randomised clinical trials, showing decreased tumour control

TABLE 7.5 Dose Reduction Factor for Normal Tissues of Interest for Head and Neck Radiotherapy after the Administration of Amifostine

Tissue	Dose Reduction Factor
Salivary gland	2.0
Oral mucosa	>1
Bone marrow	2.4–3
Immune system	1.8–3.4
Skin	2–2.4

Source: Hall EJ, *Radiobiology for the Radiologist*, 5th edition, Lippincott Williams & Wilkins, Philadelphia, 2000.

in the vitamin-treated arm (Bairati et al. 2005). The trial accrued 540 HNC patients that were randomised into two arms: one that received alpha tocopherol and beta carotene supplements during radiotherapy and a placebo arm (i.e. only radiotherapy). While acute adverse effects during radiotherapy were reduced in the arm receiving supplements (odds ratio [OR] 0.72, 95% confidence interval [CI]), the overall quality of life was not improved, and the rate of local recurrence was higher in the arm receiving supplements (hazard ratio [HR] 1.37, 95% CI) (Bairati et al. 2005), concluding that vitamin supplementation might compromise the effectiveness of radiotherapy.

The hormone *melatonin* was found to be a possible candidate as radioprotector, by increasing the expression of antioxidant enzymes such as superoxide dismutase (SOD) and glutathione peroxidase. Current research on melatonin as radioprotector is mainly focused on brain malignancies (Berk et al. 2007) and no studies on a possible effect on HNC patients have been reported so far.

Radiation mitigators, a different class of radioprotectors, are an important research topic for over a decade. They are agents that either counteract acute toxicity by interfering with the post-irradiation processes or repair tissue injuries by stimulating repopulation in the affected cell compartments (Citrin et al. 2010). Growth factors and cytokines can play the role of radiation mitigators when administered close to irradiation time. The keratinocyte growth factor (KGF) is a potent mitigator as it stimulates several critical processes in cells such as DNA repair, proliferation, and alters the cell's ability to scavenge free radicals (Lombaert et al. 2008). One KGF-based agent, named *palifermin*, was tested in both preclinical as well as clinical situations with promising results.

Studies in mice showed that palifermin was able to prevent radiation damage to salivary glands by expanding the stem/progenitor pool when administered either before or after irradiation. An interesting result is the observation whereby pretreatment with KGF was shown to enhance the absolute number of salivary glands, while post-treatment with KGF induced an accelerated growth of the surviving stem cells (Lombaert et al. 2008). This accelerated growth was more pronounced the larger the number of post-irradiation surviving stem cells was.

Palifermin was also proven effective in reducing severe mucositis in HNC patients treated with cisplatin-based chemoradiotherapy (Le et al. 2011). In a randomised placebo-controlled trial enrolling 188 patients, half of HNC patients treated with conventionally fractionated chemoradiotherapy also received palifermin (180 μg/kg) before starting treatment and then once a week for 7 weeks. The incidence of oral mucositis was notably lower in this arm as compared to placebo (54% vs 69%, p = 0.041), and the median duration of severe mucositis was also significantly reduced in the palifermin arm (5 vs 26 days). Overall survival and progression-free survival were not influenced by the radiation mitigator and they were similar between the two arms (Le et al. 2011). Weekly palifermin of 120 μg/kg was administered during the course of chemoradiotherapy in a European randomised placebo-controlled trial that enrolled a similar number of patients (186) as the above American study and reported very similar results, thus confirming the potential role of palifermin in the prevention of radiation-induced mucositis (Henke et al. 2011). Past trials evaluating smaller doses of palifermin (60 μg/kg per week) with concurrent chemoradiotherapy in HNC patients did not show any protective effect (Brizel et al. 2008). Although higher doses, as reported by the above randomised trials have proven effective, caution should be taken when designing further trials as the toxicity profile of palifermin and its long-term effects are not fully documented.

Clinical results are promising in certain areas of normal tissue protection; however, more research is needed in order to determine the definitive role of radiation protectors and mitigators in the management of HNC.

CHAPTER 8

The Treatment of Head and Neck Cancer

8.1 TREATMENT SELECTION

HNCs represent a complex group of diseases that require the involvement of a multidisciplinary team to evaluate each individual case and to consider the treatment options (Lo Nigro et al. 2017). Consequently, cancer care teams for HNCs include not only medical and radiation oncologists, but also otolaryngologists as well as oncology and reconstruction surgeons. Other specialists like dentists, speech and hearing therapists, dietitians and psychologists may also be included in some cases.

The selection of the treatment for each patient depends on many factors that include the type, stage and location of the disease, the performance status of the patient, the possible side effects and, not in the least, the preferences of the patient. It is important to recognise that although the elimination of the tumour is the primary goal of the treatment, preserving the function of the adjacent organs and tissues is almost equally important both from physiological and social perspectives. This is a particularly important aspect for HNCs that appear in an anatomical region adjoining, in a relatively small space, critical organs, blood vessels and nervous system pathways that could lead to serious side effects if damaged. Therefore, to balance tumour cure against normal tissue complications, all the main treatment approaches available for cancer management, namely surgery, radiation therapy, chemotherapy and targeted therapies are considered for HNCs, either alone or in various combinations.

Surgery is one of the main treatment options for early stage tumours, which represent about one-third of the HNCs. It aims to remove the tumour and a rim of healthy tissue surrounding it during an operation. It could also include the exploration of the adjacent lymph nodes and the possible removal of the affected ones (de Bree & Leemans 2010). However, it carries a high risk for facial disfigurement and other severe side effects.

Radiation therapy is the other main treatment option for early stage tumours and an attractive alternative to surgery when the latter is deemed too dangerous and likely to lead to severe side effects, for example because of the proximity of the tumour to critical structures like major vessels. Furthermore, radiation therapy could also be used as an adjuvant treatment in those cases in which surgery failed to achieve its objectives, i.e. the removal of the tumour and the safety margin around it (de Bree & Leemans 2010).

For locally advanced diseases, the treatment is more intensive and it involves several of the treatment approaches mentioned above in various combinations. Thus, surgery could be used for the resection of the primary tumour and affected lymph nodes, followed by locoregional radiotherapy. This has been the mainstay of treatment for many years, until results emerged in the 1990s that showed that organ preservation strategies employing neoadjuvant chemotherapy followed by radiotherapy could lead to comparable survival outcome as surgery-based treatments, but without the loss of speech, disfigurement, and other severe side effects that may result from surgery in the head and neck region (Wang & Knecht 2011). Nevertheless, other combinations of therapeutic approaches have also been used (Cohen, Lingen & Vokes 2004), for example concomitant chemoradiotherapy, postoperative chemoradiotherapy or radiotherapy given first to reduce the size of the tumour, followed by surgery and chemotherapy.

These treatment options could also be accompanied by targeted radiotherapy that, as the name suggests, targets specific genes, proteins, or even the tumour environment that are involved in tumour growth (Bozec, Peyrade & Milano 2013). For example, inhibition of the epidermal growth factor receptor (EGFR) has been shown to slow the growth of some head and neck tumours and could; therefore, contribute toward better outcomes. Nevertheless, it has to be acknowledged that not all tumours have the same targets and that extensive tests are needed to identify the proper biomarkers that are amenable to targeting. More details about these aspects could be found in the section dedicated to targeted therapies.

Very advanced cases of HNCs, which often involve a distant metastatic spread of the disease, require a rethinking of the treatment strategy from locoregional approaches to systemic approaches (Hoffmann 2012). These cases often use all the options available in the treatment arsenal to address the various levels of disease.

The available treatment approaches lead to a remission of the disease in many cases. However, for a significant fraction of the patients, especially those with locally advanced disease, there is a risk that the disease will recur after the primary treatment. The available treatment approaches must take into account both the treatment received for the primary disease and the time that has elapsed from the primary treatment. Details that have to be considered in these cases are addressed in the section dedicated to retreatment. Nevertheless, the treatment options remain essentially the same as those listed above.

8.2 TARGET DEFINITION FOR HEAD AND NECK RADIOTHERAPY

Target definition for head and neck radiotherapy is a complex and still-evolving task in the management of this disease. To unify and standardise the definition of relevant structures for radiotherapy, the International Commission on Radiation Units and Measurements proposed a unifying nomenclature that has been adopted for all radiation therapy techniques (ICRU 1993, 1999, 2004, 2007, 2010).

The primary tumour volume and the affected lymph nodes are usually delineated based on clinical and pathological examinations. Traditionally, these examinations included palpations, endoscopic examinations, and biopsy results. Later years have seen the introduction of computed tomography (CT) scans, which is now the backbone of the treatment planning process. More recent years have also seen a gradual introduction into treatment planning of increasingly advanced morphological and functional imaging methods for better target definition, like MRI and PET, which established the concept of image guided radiotherapy (IGRT). For example, it is currently the routine in many centres to include into the tumour volume metabolically active regions detected by PET with suitable tracers, which may otherwise not be visible in morphological images. Other imaging methods might highlight other volumes or processes of interest.

Irrespective of the method used for characterisation, the primary tumour volume and the affected lymph nodes define the Gross Tumour Volume (GTV) according to the ICRU nomenclature. The GTV could sometimes be divided into two operational volumes, separating the GTV of the primary tumour (GTV_{tumour}) from the GTV of the affected nodes (GTV_{nodes}). The GTV is then expanded with a suitable margin to the clinical target volume (CTV) to account for microscopic infiltration of the disease beyond the margins of the GTV. HNCs have been shown to spread easily through the lymphatic system and, therefore, lymph node regions that could harbour microscopic disease are electively included into a $CTV_{elective}$. Recent years have seen the development of atlases to be used in the delineation of the lymph node regions for head and neck radiotherapy.

The distinction between the CTV encompassing the primary tumour and the affected nodes and the $CTV_{elective}$ is not only operational, the two regions being prescribed different doses, with the former receiving an extra boost dose in comparison to the latter, which led to it being often called the CTV_{boost}. As is normal practice in radiotherapy, each CTV is further expanded to account for internal motion and setup uncertainties to create corresponding planning target volumes (PTV), for example $PTV_{elective}$ and PTV_{boost}. In addition to these volumes that are part of the ICRU recommendations, Ling et al. (2000) proposed defining a series of other biological target volumes (BTV) encompassing biological processes of interest with predictive power for the evolution of the tumour. However, the concept is still under development and work is being conducted on prescription approaches to account for the targeted biological processes identified (Toma-Dasu & Dasu 2015).

Besides the tumour and target volumes, critical structures are also delineated as organs at risk (OAR) to monitor the doses they receive during the treatment. The volumes of OAR could also be expanded to account for internal motion and setup uncertainties leading to the delineation of planned risk volumes (PRV) for each normal structure of interest. The critical structures that are taken into account depend on the localisation of the tumour, but they typically include spinal cord, salivary glands, eyes and eye lenses, optic chiasm, brain, brain stem, pituitary gland, larynx, oesophagus, inner ears, mucosa in the oral cavity, thyroid, jaw bones, or mastication muscles.

The target and risk volumes for each patient are then used in a treatment planning and optimisation process aiming to design a plan that could deliver the prescribed dose to the target volume while minimising the doses to the OAR or at least keeping them within acceptable levels. Successful plans are evaluated based on their ability to ensure the coverage of the PTV with a certain isodose, typically 95% of the prescribed dose, while not exceeding some threshold values for the dosimetric parameters for the organs at risk. This target coverage approach ensures that the CTV receives the prescribed dose when suitable margins are used to account for internal motion and setup uncertainties. However, recent years have seen an increase in robust and probabilistic planning aiming to ensure suitable coverage of the CTV by increasing the robustness of the plan to motion and setup uncertainties, as well as other scenarios regarding tumour and organ motions. While this approach no longer needs a PTV, the PTV is still used for treatment plan comparison and dose reporting.

8.3 TREATMENT TECHNIQUES

Several radiation therapy techniques are available to deliver treatments to the head and neck region and they will be reviewed in this section. Choosing one among the relevant treatment techniques takes into account a number of aspects that include tumour shape and location, availability of technique or local customs and traditions and, not least of all, the cost of the treatment. Nevertheless, with each treatment technique, the objective is the same, to deliver the prescribed dose to the target while minimising the irradiation of the normal structures.

8.3.1 External Beam Radiotherapy

External beam photon radiotherapy is the most common radiotherapy technique used for head and neck treatments. It involves the irradiation of the target using an external radiation source. Nowadays, radiation is commonly produced with a linear accelerator, but irradiation with natural radioactive sources of Cs-137 and Co-60 has been extensively used in the past. Furthermore, several techniques are available for EBRT, each with various degrees of complexity.

Nowadays, the choice of radiation treatment lies mainly between three-dimensional conformal radiation therapy (3D-CRT), intensity modulated radiation therapy (IMRT) with fixed beam angles or volumetrically modulated arc therapy (VMAT). The former technique uses multileaf collimators (MLC) as well as dynamic wedges to create suitable dose distributions typically from opposed, fixed angle, oblique fields. It is quite a robust technique that is useful for small or convex-shaped, unilateral targets. In contrast, IMRT with fixed beam angles or rotational therapy are later developments in irradiation techniques that use MLC to modulate the photon fluence across the fields aiming to deliver complex three-dimensional dose distributions with steep gradients that could better conform the high value isodoses to the target volumes. This creates the premises for better normal tissue sparing, especially in cases of large or concave targets, as is for example the case of bilateral targets with the spinal cord in the posterior concavity. Another difference between these techniques is that in 3D-CRT the fields deliver dose concurrently to all the target volume, while in IMRT there is a temporal variation in the irradiation of the different parts of the target. An alternative technique, not as widespread as the other two, is CyberKnife that makes use of a linear accelerator mounted on a robotic arm to deliver pencil beams from a large number of directions to cover the target. While the technique is largely used in stereotactic treatments, reports exist regarding its use also for boosting the dose in head and neck radiotherapy.

Irrespective of the external beam irradiation technique used, the patient is immobilised with the help of a mask. Typical for head and neck radiotherapy are 5-point thermoplastic masks that immobilise the head, neck and shoulders of the patient. The aim is that the mask ensures a close fixation of the head of the patient that minimises rotational movements around both frontal and axial axes for the whole duration of the treatment. It is especially important for head and neck patients to check that the mask does not became loose during the treatment as the patient could lose weight. Similarly, tumour growth or inflammations could increase the tightness of the mask which will make it more uncomfortable for the patient. Both these changes could lead to changes in the positioning of the patient relative to the planning stage that would translate in changes to the delivered dose distributions.

Treatment planning follows the usual practice in radiation therapy, with most of the treatments employing a coplanar disposition of the fields. 3D-CRT employs isocentric, opposing oblique fields, typically with the isocentre placed near the centre of the target volume. Dynamic wedges and additional fields could be used for dose homogenisation across the target. Typically, low photon energy, e.g. 6 MV, is used for treatments in the head and neck area, although higher energies might also be considered for the additional fields. Dose distributions in 3D-CRT are characterised by high lateral gradients, with the volume of tissue traversed by the radiation fields receiving close to prescription dose and those outside it receiving mainly scattered radiation. IMRT with fixed beams typically uses an odd number of fields equally spaced in terms of angles, while rotation therapy approaches use one or several fields that are continuously rotated around the patient. Usually, low photon energies are used, both due to the low depth involved and the improved penumbras that could be achieved with these energies. In rotational therapy, arc segments that imply the irradiation of the target through bony or thick structures, e.g. through the shoulders, are blocked to avoid unnecessary radiation burden to the normal structures. The fluence from each beam is then optimised to lead to the required balance between target coverage and normal tissue burden. This is usually achieved through an iterative approach in which the dose contribution from each beam segment is compared with the optimisation objectives and the weight of each element is modified according to the determined deviation from the objectives. The result is rather heterogeneous beam fluence from each field that is achieved modulating the position of the MLC, either in discrete steps as in the step-and-shoot techniques or dynamically as in the sliding window techniques. Later years have seen the frequent use of rotational therapies, either VMAT or tomotherapy, that use the simultaneous changing of the leaf patterns while the source rotates around the patient. Multiple arcs could be used to improve the dose gradients and, therefore, the achievable dose levels.

Various levels of dose prescription exist for EBRT, depending on the type and stage of the tumour or the role of radiotherapy in the management of the disease. Considering conventional fractionation with daily fractions of 2 Gy, typical doses prescribed to the elective target volume range from 46 to 50 Gy, followed by additional 20–24 Gy prescribed to the boost target volume. The alternative is a so-called simultaneous integrated boost which aims to deliver both dose levels in the same treatment session, for example a fractional dose of 2 Gy to the elective volume and a fractional dose of 2.2 Gy to the primary tumour for the same number of fractions. It is noteworthy that many other fractionation schedules have been tried with external beam photon radiotherapy to apply the fractionation principles known as the Rs of radiotherapy and the reader is directed to reviews on the topic (Bourhis et al. 2006). A summary of some of the treatment schedules employed in the radiotherapy of HNC ranging from hyperfractionated to escalated doses per fraction together with the reported rates of local control is given in Table 8.1.

Analyses of clinical studies have highlighted the impact of various radiobiological factors on the local control from radiotherapy. Thus, it has been suggested that tumour volume is one of the main predictors of radiotherapy outcomes, as larger tumours contain more tumour cells than smaller tumours. Assuming that the number of tumour cells is proportional to the tumour volume, the ratio of tumour control probabilities from the same treatment for tumours with different volumes is in an exponential relationship with the ratio of volumes (Equation 8.1):

$$TCP_2 = TCP_1 \wedge (V_2/V_1) \tag{8.1}$$

where TCP_1 and TCP_2 are the tumour control probabilities of tumours with volumes V_1 and V_2, respectively (Brenner 1993). Clinical volume response data generally agree with

TABLE 8.1   A Summary of the Main Treatment Schedules Employed in the Radiotherapy of HNC Specifying the Number of Fractions in Which the Treatment is Delivered, the Dose Per Fraction, the Overall Treatment Time (OTT) and the Reported Rates of Local Control

Schedule	No. of fx	Dose per fx (Gy)	OTT (days)	Local Control	Reference
Conventional	35	2	49	46%	Horiot et al. (1997)
EORTC 22851	45	1.6	35	59%	Horiot et al. (1997)
Conventional	33	2	46	50%	Dische et al. (1997)
CHART	36	1.5	12	54%	Dische et al. (1997)
Conventional	35	2	49	47%	Poulsen et al. (2001)
TROG	33	1.8	24	52%	Poulsen et al. (2001)
DAHANCA long	34	2	46	60%	Overgaard et al. (2003)
DAHANCA short	34	2	39	70%	Overgaard et al. (2003)
CAIR-1 short	38	1.89	38	75%	Skladowski et al. (2006)
CAIR-1 long	38	1.89	53	33%	Skladowski et al. (2006)
Escalated	20	2.55	25	51%	Cummings et al. (2007)
Hyperfractionated	40	1.45	25	59%	Cummings et al. (2007)
GORTEC 99-02	36	1.8	24	53%	Bourhis et al. (2011)

the relationship predicted by Equation 8.1, but with a less steep dependence, suggesting that other factors might also influence local control and, therefore, act as sources of heterogeneity (Bentzen & Thames 1996, Dubben, Thames & Beck–Bornholdt 1998). Indeed, proliferation parameters and intrinsic radiosensitivity have also been found to have predicting power for local control from radiotherapy (Dubben, Thames & Beck–Bornholdt 1998). Furthermore, functional imaging has highlighted yet other factors that influence local control, such as metabolic activity (Seol et al. 2010) or hypoxia (Dunst et al. 2003). Nevertheless, primary tumour volume has a strong impact on local control after definitive radiotherapy for HNC (Mendenhall et al. 2014), but also for combined treatment modalities (Knegjens et al. 2011).

Besides these tumour intrinsic aspects, overall treatment time (OTT) was also found to strongly influence local control for HNC due to the deleterious effect of accelerated repopulation of tumour clonogens (Bese, Hendry & Jeremic 2007). Consequently, interruptions that affect the programmed OTT need to be compensated for through the delivery of extra treatment fractions, the delivery of several fractions per day to maintain the programmed time-course for irradiation or other manipulations of the fractionation schedule. The reader is directed to the appendix on radiobiological modelling for calculations of isoeffectiveness that may be used to compensate for gaps in head and neck treatments.

8.3.2 Brachytherapy

Brachytherapy is a radiation therapy technique using sealed radiation sources placed within or adjacent to the treated area, to achieve a high dose differential between the irradiated volume and the surrounding normal tissues. Brachytherapy is used less than EBRT in the treatment of HNCs, as it is quite labour intensive, but it can provide an effective treatment in some situations. The abrupt dose fall off that characterises brachytherapy indicates it for well localised tumours that are very close to critical structures, as well as for treating recurrence cases when critical structures have reached their tolerances and no further external radiotherapy could be given. This latter case is also known as salvage radiotherapy. In the case of head and neck treatments, brachytherapy could be used either alone or in combination with EBRT.

For HNCs, brachytherapy could be delivered either with interstitial implants or with intracavitary implants. Interstitial brachytherapy is delivered through catheters inserted through the tumour tissue, which require access to an operating room and general anaesthesia for the patient. Several individual treatments could be delivered over several days during which the patient is hospitalised. In contrast, intracavitary brachytherapy is delivered through catheters inserted in the oral or nasal cavity to access the tumour site. This is a much simpler form of brachytherapy which could also be delivered over several days, but it requires only local anaesthesia and could be delivered in an outpatient setting.

A variety of source types and isotopes are available for brachytherapy. Historically, Radium needles were the first brachytherapy sources that were directly loaded into the tumour bed. These were later replaced by other isotopes, as was the direct loading technique replaced by afterloading technique, either manual or remote. Currently, brachytherapy for HNCs is delivered using the remote afterloading technique using Ir-192 sources, although both Co-60 and Cs-137 sources are also available. In the afterloading technique, catheters are placed about 1–2 cm apart in the tissue in the first step of the procedure and are then connected to the afterloading machine for the remotely controlled insertion of the sources during the treatment step. The insertion of the sources takes place according to a pre-calculated plan that optimises the position of the radioactive source in the catheters to ensure a uniform dose to the tumour bed and to minimise the irradiation of the adjacent structures.

Brachytherapy is usually delivered with HDR, typically exceeding 12 Gy/h, although lower dose rates (LDR) have also been historically used. This has prompted the development of irradiation techniques where irradiation is delivered in relatively short HDR pulses of a few minutes separated by intervals of several hours to allow for repair of sublethal damage.

This technique is referred to as pulsed dose rate (PDR) brachytherapy.

The Groupe Européen de Curiethérapie and the European SocieTy for Radiotherapy & Oncology (GEC-ESTRO) have recently issued a set of recommendations for brachytherapy for head and neck squamous including both primary and recurrent squamous cell carcinoma of the head and neck. The lack of randomised trials on brachytherapy of HNC, however, made difficult the formulation of the present consensus recommendations for LDR, PDR and HDR brachytherapy issued by GEC-ESTRO (Mazeron et al. 2009). The recommendations include a comprehensive pallet of clinical aspects concerning patient selection, treatment strategy, target definition, implant techniques, dose and dose rate prescription, treatment planning and reporting, treatment monitoring, catheter removal and post-treatment patient care and follow-up per tumour site lip, oral mucosa, mobile tongue, floor of mouth, oropharynx, nasopharynx, paranasal sinuses (Mazeron et al. 2009). A follow-up of the initial recommendations was recently issued by the same forum (Kovács et al. 2017).

Common dose regimens for different treatment sites when brachytherapy is used as the primary treatment modality specify 15–30 Gy as a brachytherapy boost after 46–50 Gy of EBRT for the most common indication in advanced cases of oral cavity cancer or 70 Gy as LDR over 4–9 days or 60 Gy in 10 fractions as HDR when brachytherapy is used alone. In case of oropharynx cancer, brachytherapy is usually combined with EBRT. The prescribed doses, the dose regimens and the technique used for delivering the treatment highly depend on the exact location of the tumour and proximity of the sensitive OAR for the oropharynx, nasopharyngeal and head and neck superficial cancer (Kovács et al. 2017).

8.3.3 Proton and Heavy Ion Therapy

Radiation therapy employing charged particles like protons and heavier ions is a form of external beam radiation therapy that provides a dosimetric advantage over the photon techniques due to the interaction properties of the charged particles. Thus, charged particles have a finite range in tissue and deposit most of their energy in the Bragg peak in the distal part of their range. Furthermore, the scattering of particles, especially heavier ions, is much reduced leading to sharp penumbras that translate into increased sparing of out-of-field organs. These features allow a more favourable dose deposition in tissues compared to photons, both with reduced dose deposition in the entrance channel and virtually zero dose deposition beyond the Bragg peak. The principles of particle therapy have been outlined quite early (Wilson 1946), but the technique has slowly been adopted clinically, mainly due to the limited availability of the equipment required to accelerate the particles to the energies needed. Only recent years have seen an increased proliferation of particle therapy centres, but the number of patients treated with this modality is still very low (Jermann 2015).

Particle therapy could now be delivered with either of two techniques. The first one to be used chronologically is passive scattering when a narrow particle beam is scattered laterally and modulated in energy to ensure the simultaneous coverage of the target. The technique requires the use of compensators to ensure the distal conformity of the dose distribution and collimators to allow for the lateral conformity of the beam to the target. These modifiers are field-specific and require individual design, manufacturing and positioning in the therapeutic beams. The alternative technique that has gained increasing ground is active scanning that uses magnetic fields to scan laterally the narrow particle beams and active energy modification for changing the energy of the particles and, therefore, their penetration depth. The technique does not require beam-specific modifiers, but the lateral penumbras that could be achieved may be somewhat inferior to those that could be obtained from the use of collimators.

The clinical experience with respect to the use of particles for radiation therapy is small in comparison to photon therapy and; therefore, dose prescription and plan evaluation in particle therapy is largely based on photon experience, taking into account the relative biological effectiveness (RBE) of the particles that is higher compared to photon radiation. This extrapolation of knowledge is further hampered by the uncertainties with respect to the RBE of the protons (Dasu & Toma-Dasu 2013; Jones 2016) and of the heavier ions (Kase et al. 2008) and, therefore, the exploration of the potential of this form of radiotherapy is pursued with caution.

Few clinical studies exist on particle therapy for HNCs in comparison to photons, but the number of trials is increasing rapidly (Leeman et al. 2017). The clinical studies have been complemented by a number of treatment planning studies investigating the theoretical potential of particle therapy. These studies have shown that particle therapy is particularly favourable for skull base tumours as well as for orbital, periorbital and paranasal tumours as the limited range of the particles could spare such critical structures as the brain stem, eyes and optic nerves. For other types of head and neck tumours, particle therapy could lower the dose to the salivary glands, thus lowering the risk for xerostomia. However, the potential of particle therapy is reduced for laryngeal cancer as well as for targets in or close to the skin (Ask et al. 2005). Another area of potential advantages of particle therapy is the treatment of recurrent and second primary HNCs in case where the position of the recurrent disease in relation to previously irradiated structures deem the patients unsuitable for further external photon beam irradiation. From this point of view, particle therapy could be regarded as an alternative to salvation brachytherapy. Nevertheless, these are only broad term indications and the potential for improvement that could be achieved from particle therapy has to be investigated on individual basis for each patient.

The limited range of particles in tissues, besides providing a dosimetric advantage, also poses further requirements with respect to the movable and sometimes highly heterogeneous structures in the head and neck region for ensuring the relative position of the target to the treatment beam and the normal tissues at risk. This is because the range of the particles and the position of their Bragg peak are strongly dependent on the composition and thickness of tissue along their path length. A useful concept to quantify the penetration of particles in tissues is the water equivalent thickness (WET) along the path of the particles. From this perspective, it is important to recognise that increases of the WET lead to limited penetration of the particles that in turn result in underdosage of the distal part of the target. In contrast, decreases of the tissue WET would lead to increased penetration of the particles which would lead to both underdosage in the proximal part of the target and overdosage beyond the distal part of the target, which could be critical for some normal tissues at the limit of their tolerance. These effects are significantly more important for particle therapy than for photon therapy, where changes in tissue heterogeneity result in comparatively small dosimetric changes. For the heterogeneous head and neck region, changes in WET along the particle track could be caused by a large number of reasons. Thus, changes in patient setup could easily change the tissue composition along the particle beams. Changes in WET could also be brought by modification of the patient contour either from inflammations and tumour progression or from weight loss, but also from changes in mask position, especially in case of beam openings, or variations in folds of the patient skin. Furthermore, filling or emptying of cavities, e.g. sinuses, should be carefully monitored during the course of the treatment to ensure the situation at planning. All these reasons demand increased use of imaging for plan verification in the case of particle therapy.

8.3.4 Combined Treatment Modalities

Randomised trials have shown that chemoradiotherapy, particularly cisplatin-based chemoradiotherapy improves survival compared to radiation as a sole agent, reason why this combined treatment represents the standard of care in locally advanced HNC. When combined with radiation cisplatin is a potent chemotherapeutic agent through a variety of mechanisms, including inhibition of DNA repair, cell-cycle arrest (Lawrence, Blackstock & Mc 2003), radiosensitisation via DNA adduct formation and inhibition of angiogenesis (Yoshikawa, Saura & Matsubara 1997).

Nevertheless, cisplatin-induced normal tissue toxicity is often a dose-limiting factor. Side effects such as nephrotoxicity, ototoxicity, neurotoxicity and haematological toxicities are very common with the administration of cisplatin. Since the platinum family has proven efficacy in head and neck carcinomas, the aim was to develop platinum compounds that are less toxic to the normal tissue (see also Table 8.2). Therefore, the next generation of platinum-based agents was represented by carboplatin. Studies comparing the effect of cisplatin and carboplatin on head and neck malignancies suggest that carboplatin is inferior to cisplatin regarding tumour control (Go & Adjei 1999). While carboplatin has replaced cisplatin in the management of certain neoplasms (such as ovarian cancers), its success in HNCs is limited. Oxaliplatin is a third-generation agent of the platinum family. This drug exhibits radiosensitising properties similar to cisplatin, but with reduced toxicity (Espinosa et al. 2005; Stordal, Pavlakis & Davey 2007). Yet, to date, oxaliplatin is more

TABLE 8.2   Platinum Compounds as Chemotherapeutic Agents for HNC with Their Possible Advantages and Drawbacks Compared with Cisplatin

Platinum Compound	Benefits	Drawbacks
Cisplatin	Potent cytotoxicity Most trialled agent	High normal tissue toxicity. Drug resistance
Carboplatin	Reduced normal tissue toxicity (no nephrotoxicity).	Inferior tumour response rate. Myelosuppression
Oxaliplatin	Reduced normal tissue toxicity and better tolerability. Greater cytotoxicity and inhibition of DNA synthesis.	Neurotoxicity. Conflicting results on the efficacy on cisplatin-resistant cell lines.
Nedaplatin	Reduced nephrotoxicity and gastrointestinal toxicity. Similar tumour control.	Often exhibits cross-resistance with cisplatin thus its clinical application is limited.
Mitaplatin	Exhibits toxic effects on cisplatin-resistant head and neck tumour cells. Better selectivity for tumour cells than cisplatin.	More research is needed to prove its clinical efficacy.
Enloplatin	Tested in the 1990s without successful clinical implementation	
Tetraplatin/ormaplatin	Under investigation	

frequently used for colorectal cancers (Pandor et al. 2006) than for squamous cell carcinomas showing that more convincing evidence is needed for its clinical implementation. Nedaplatin was developed to reduce nephro- and gastrointestinal toxicities that are commonly acquired during cisplatin-based chemotherapy. This drug is more popular in some Asian countries where nedaplatin-based clinical trials on HNC (mainly nasopharyngeal carcinoma) have successfully been conducted (Peng et al. 2013, Xu et al. 2014, Yin et al. 2015). Given the fact that nedaplatin often exhibits cross-resistance with cisplatin, its clinical use is limited (Liu et al. 2015).

Mitaplatin, the newest member of the platinum family, is a fusion between cisplatin and the orphan drug dichloroacetate previously developed for lactic acidosis (Dhar & Lippard 2009). Beside the fact that mitaplatin was shown to exhibit toxic effects on cisplatin-resistant head and neck tumour cells, the drug has a better selectivity for tumour cells than cisplatin, which has the potential to reduce normal tissue toxicity (Xue et al. 2012).

Cisplatin is commonly administered in cocktail combination with different classes of drugs for better radiosensitisation and also in order to overcome drug resistance. As shown in Chapter 2, some of the most commonly used drug classes are the antimetabolites (5-fluorouracil), antibiotics (mitomycin-C, bleomycin) and taxanes (paclitaxel). While different drug classes exhibit different cytotoxic mechanisms (see Table 2.2), the interaction of these drugs with radiation enhances the antitumour effect of radiotherapy.

Currently, the optimal treatment regimen for locoregionally advanced HNC is the concurrent administration of chemo- and radiotherapy. While chemotherapy has also been trialled as induction and/or adjuvant treatment, in HNC the current drug combinations were shown to be ineffective (Milas et al. 2003). However, given the heterogeneous design of various treatment regimens, it is difficult to define the optimal multi-agent chemotherapy for these cancers (Browman et al. 2001).

Regarding the radiotherapy component, IMRT is becoming the dominant technique in HNC irradiation (Ang et al. 2012). The advantage of IMRT over 3D-CRT consists of a better management afforded by IMRT, whereby with each fraction, different doses are delivered to the high- and low-risk tumour volumes using a simultaneously integrated boost technique. With 3D-CRT, the two volumes are usually dealt with differently, with the low-risk volume receiving treatment for 5 weeks while the high-risk volume receives treatment for 7 weeks. The PARSPORT multicentre randomised controlled trial is a fitting example of the efficacy of IMRT over conventional radiotherapy (Nutting et al. 2011). One of the greatest achievements of IMRT is the sparing of the parotid gland, which in turn reduces the incidence of xerostomia, the most common toxicity after head and neck radiotherapy. When combined with chemotherapy, IMRT was shown again to lead to superior results compared with conventional chemoradiotherapy, in both HPV+ and HPV– tumours (Studer et al. 2010; Montejo et al. 2011; Clavel et al. 2012). The more conformal dose distribution with manageable late toxicities confers IMRT a higher status than conventional irradiation.

Despite its normal tissue toxicity, cisplatin is the most reported chemotherapeutic agent in the head and neck literature, fact that makes this drug one of the first treatment choices (Ang et al. 2012). IMRT combined with cisplatin-based chemotherapy is currently the most advocated treatment for locoregionally advanced squamous cell carcinomas of the head and neck.

8.4 TREATMENT RESPONSE MONITORING AND ADAPTATION

Head and neck tumours have limited spread patterns and; therefore, achieving local control is essential for curing the disease. Indeed, one of the most quoted reasons for treatment failure for locally advanced squamous cell carcinoma of the head and neck (HNSCC) is poor locoregional control (Argiris et al. 2008; Bussink et al. 2010).

Recognising the importance of objective monitoring of local control, the European Organisation for Research and Treatment of Cancer (EORTC) has proposed a set of criteria to assess the response to anti-cancer treatment in solid tumours (Eisenhauer et al. 2009). These are known as the Response Evaluation Criteria in Solid Tumours (RECIST) set of rules and are among the most commonly used criteria for response evaluation. They are based on measurements of all the measurable lesions that could be identified before the start of the treatment. According to this set of rules, the sum of the longest diameter (LD) for all target lesions will be calculated to be used as reference to characterise the objective tumour response after the treatment and even as the treatment progresses. Consequently, target lesions could be classified in the following categories:

- *Complete Response (CR)*, when all target lesions have disappeared

- *Partial Response (PR)*, when the sum of the LD of target lesions shows at least a 30% decrease compared with the baseline sum LD

- *Stable Disease (SD)*, when neither sufficient shrinkage to qualify for PR nor sufficient increase to qualify for progressive disease has been seen

- *Progressive Disease (PD)*, when at least a 20% increase in the sum of the LD of target lesions is seen compared with the baseline sum LD or when one or more new lesions have appeared

These criteria are usually assessed based on morphology information from CT or MRI. Nevertheless, morphological imaging has some limitations as it could give for example false indications of PD when scar tissue is formed. This, together with recognising that functional changes typically precede changes in anatomy (Bussink et al. 2010), have indicated functional imaging as a more sensitive tool for treatment response monitoring. Thus, functional imaging could be used for post-treatment disease assessment, but the options available for poorly responding patients are generally limited to salvage surgery (Bhatnagar et al. 2013).

Increasing the local control for head and neck treatment is not straightforward as the balance with complication rates is very delicate and normal tissue toxicity is usually 'pushed to what must be the limits of human tolerance' (Corry, Peters & Rischin 2010). Therefore, proposals have been made to adapt the treatment to the features that confer resistance to the tumours. Three main mechanisms have been identified as responsible for resistance to radiotherapy for HNCs: hypoxia, accelerated repopulation and intrinsic radioresistance (Bussink et al. 2010). Consequently, imaging techniques have been sought to monitor these aspects to help identify the patients with resistant tumours that might be in need of dedicated approaches. Adapting the treatment to pre-treatment imaging features is generally included in the IGRT approach for treatment individualisation.

Nevertheless, treatment for head and neck generally proceeds for several weeks without taking into account the changes that might occur during this time. However, the morphology of the patient and the tissue function could change both directly and indirectly due to the treatment and the disease itself. Thus, radiation could induce both tumour shrinkage and tissue inflammations, while the patients could both go up and down in weight during the course of treatment. These changes might affect the positioning of the targets and the surrounding normal tissues in relation to the treatment fields and could result in both an underdosage of the target and an overdosage of the normal tissues. Furthermore, the functional features of tumours and normal tissues might also change, even when morphological changes are not obvious. Consequently, proposals have been made to monitor the response of the patient and to adapt the treatment to the changes that might occur in the target and the surrounding normal tissues. This approach has been termed treatment adaptation and is still a developing field for tumours of the head and neck as well as for other sites.

Until recent years, the focus of treatment response monitoring overlapped to a large extent with the aim of IGRT for HNCs, namely the monitoring of the morphological changes of the tumour as well as the changes in size, shape and position of the OAR and the normal tissue in general. However, with the development and the increased clinical availability of the functional imaging techniques, new facets have started to be added to the monitoring treatment response with special emphasis on early changes during the treatment (Gregoire & Chiti 2011; Castaldi et al. 2012). Indeed, the very aspects that make the object of targeting in IGRT could also be monitored during the treatment and correlations could be sought between changes in these parameters and treatment outcome. Therefore, treatment response monitoring could potentially improve the outcome of the treatment by allowing the identification of the patients that should benefit from individualised treatment adaptation.

Treatment monitoring based on functional imaging for treatment adaptation requires a series of tools for its successful implementation. Thus, image coregistration is essential for treatment monitoring, as is for image-based treatment planning. Rigid image coregistration is now available in many treatment planning systems, but deformable image registration is being increasingly used to account for deformation of structures. Nevertheless, to minimise the errors that might appear due to deformable registration, it is recommended that functional imaging is acquired in radiotherapy treatment position, e.g. by using flat couch tops. For MRI, it is essential that MR-compatible immobilisation devices are used. Also, the use of additional coils must not interfere with the surface of the patient, the latter aspect being especially important for particle therapy due to the finite range of the particles in tissue.

Repeated imaging during the treatment produces a large amount of data that requires additional processing. The amount of work required with respect to structure delineation and analysis has increased interest in automatic processing tools, such as automatic structure delineation and propagation, automatic analysis and even automatic treatment planning. While these will be useful tools, considerable work is still required for their validation, and; therefore, visual interpretation of images by skilled experts is still the norm for response monitoring and treatment adaptation.

Several key issues are currently under investigation in relation to functional response monitoring. One of these is the choice of functional imaging modality. So far, PET imaging has been the most widely used modality and criteria have been provided for the quantitative assessment of the metabolic tumour response, paralleling the RECIST set of rules (Wahl et al. 2009). While most of the existing studies have focused on the potential of the global standard uptake values of the investigated tracer (Troost et al. 2010; Sun et al. 2012), more advanced methods have also been explored (Toma-Dasu et al. 2015). The potential of MRI for response monitoring has also been investigated (Dirix et al. 2009; Kim et al. 2009), although there are fewer studies than for PET. It has to be mentioned that various imaging modalities and tracers offer information on the different facets of tumour resistance and; therefore, they must be regarded as complementary rather than replacements.

The dynamics of the investigated aspects is another key issue for response assessment. For example, tumour hypoxia has been shown to have a dynamic behaviour even in the absence of treatment (Cardenas-Navia et al. 2008) and; therefore, variations in hypoxic patterns may not necessarily reflect the effect of the treatment. Furthermore, this dynamic behaviour advises against treatment adaptation strategies based on dose redistribution, i.e., the intentional decrease of dose to regions not showing hypoxia, as its dynamic evolution could lead to later mismatches between hypoxic regions and dose hotspots which might eventually lead to treatment failure.

Another key issue for response assessment is the optimal time window for imaging. Very early imaging could fail to highlight relevant changes for treatment outcome, while late imaging may not allow enough time for treatment adaptation and could also be subject of treatment-related confounding factors. Thus, it has been postulated that imaging one or two weeks into the treatment would provide the necessary balance between the relevance of the imaging results and the time left for treatment adaptation (Toma-Dasu & Dasu 2015), but further validation work is needed.

Last, but definitely not the least, is the establishment of the metric to be used for treatment adaptation, going from changes in image parameters to prescribed dose. Many studies have used a linear relationship for dose conversion, but this approach is only an empirical solution, since it does not reflect the complexity of biological processes. From this perspective, nonlinear approaches taking into account the mechanistic details of the relationships might be better suited for treatment adaptation (Toma-Dasu et al. 2009).

Although much work remains to be done for the exploration of response monitoring and treatment adaptation, functional imaging appears to be a promising tool that complements clinical examination and morphological imaging for HNCs (Bhatnagar et al. 2013). Furthermore, functional imaging could also be used for the objective assessment of normal tissue function and treatment adaptations could also be introduced to maintain or even improve the quality of life of the patients.

8.5 TREATMENT FOLLOW-UP

An important aspect of the management of HNC treatment including radiotherapy is the post-treatment surveillance and the corresponding optimal follow-up strategy. Although there are several guidelines issued at national and international levels regarding the follow-up of the HNC treatment, the complexity of the clinical situations in combination with pragmatic aspects lead to lack of general consensus between clinics and countries.

A comprehensive effort, however, on designing an optimal follow-up program in HNC management from both the clinical and the economical point of view has been recently published (De Felice et al. 2017) including clear guidelines for the time intervals for clinical and imaging investigations post HNCs treatment. Thus, for clinically evaluable primary tumours, such as oral cavity and lip, oropharynx, hypopharynx, larynx, nasopharynx and occult primary tumours, the first clinical examination should be performed at one month after the end of the treatment followed by examinations every 3 months during the first 2 years of follow-up. The frequency of the clinical investigations would then decrease to one every 6 months up to 5 years after the treatment. In parallel, imaging investigations are recommended 3 and 9 months post-treatment. In case of no clinically evaluable primary tumour such as paranasal sinus, salivary gland and mucosal melanoma, the same schedule is recommended for the clinical examinations, but substantially more imaging investigations are needed at 3, 9, 15, 21, 30, 42, 54 months post-treatment (De Felice et al. 2017).

CHAPTER 9

Targeted Therapies

9.1 RATIONALE FOR TARGETED THERAPIES AND ASSOCIATED CHALLENGES

As shown in the previous chapters, chemoradiotherapy is the standard treatment for locally advanced HNC. However, a major disadvantage of this approach is the increased normal tissue toxicity, particularly to the lining of the upper aero-digestive tract and the unavoidable systemic toxicity characteristic to chemotherapy. Classical chemotherapeutic agents act specifically on cycling cells, although they lack selectivity. These cytotoxic drugs do not differentiate between healthy and malignant cells, hence the commonly known side effects. To limit normal tissue toxicities, more effective targeting agents that exhibit higher specificity and selectivity are needed.

Developments in molecular biology of squamous cell carcinomas of the head and neck have revealed several molecular characteristics and pathways that provide valuable information regarding tumour progression, metastatic potential, and response to treatment. Over the last decade, the field of molecular targeted therapies for HNC has advanced, with several agents being tested clinically. These new therapeutic agents target various signalling pathways in order to interfere with tumour progression (Bozec, Peyrade & Milano 2013; Dorsey & Agulnik 2013). As such, most of the current focus is oriented toward the EGFR and the VEGF, two stimulants of tumour growth and proliferation. So far, targeting agents were designed as (1) *monoclonal antibodies* (mAb) that target specific antigens found on the cell surface or extracellular growth factors or (2) *small molecules*, such as tyrosine kinase inhibitors (TKI), that have the ability to penetrate the cell membrane to reach the target and to interfere with its enzymatic activity. EGFR is the first molecule in cancer research against which monoclonal antibodies have been developed (Mendelsohn 1997).

Nevertheless, the development of new agents for cancer treatment is rarely without challenges, as several factors come into play: target identification and validation, the need for high tumour specificity, selective targeting of malignant versus normal cells, optimal delivery, the lifetime of the targeting complex in vivo for its effective action, resistance to treatment, and so on.

Whether part of the cancer cell (such as EGFR) or of the tumour microenvironment (such as VEGF), targets for cancer treatment need to be identified and validated. Some of the molecules identified as targets may be present in healthy tissues, but they are often overexpressed or mutated in malignant cells.

Differential targeting and inhibition of various molecular signalling pathways in malignant and healthy cells is a critical requirement for a drug intended for cancer treatment. These distinct effects must be tested in pre-clinical settings in order to assess and confirm differential efficacy and toxicity (Macha et al. 2017). Often though, in situations when targeting molecules are present in healthy cells and tumours alike, inhibition of the molecule might result in adverse events. This is the case with EGFR that is present in the skin and mucosa, thus, the inhibition of EGFR can result in skin rash and gastrointestinal toxicities.

Furthermore, since agents designed for targeted therapy are often administered concurrently with radiation or chemotherapy, it is desirable for these agents to exhibit radio- or chemosensitising properties. This effect can be achieved through inhibition of radiation/chemotherapy-induced DNA repair, induction of apoptosis, targeting, and inhibiting the subpopulation of cancer stem cells (Macha et al. 2017).

As resistance to single agents is a common reason for treatment failure, it is likely that a combination of agents will be required to keep the balance between tumour control and normal tissue toxicity. This was the case of the initially promising trastuzumab, a monoclonal antibody against the epidermal growth factor receptor HER 2, in the treatment of metastatic breast cancer (Jones & Budzar 2009) and of imatinib, a TKI that targets the oncogenic protein BCR-ABL expressed at high levels in chronic myeloid leukaemia (Gorre et al. 2001).

Accurate patient selection is a critical factor that dictates the success of targeted therapies. Gene microarray analysis potentially allows prediction of tumour response to treatment, thus foreshowing a future in personalised medicine (West, Elliott & Burnet 2007). Gene expression analysis of the tumour samples originating from HNC patients treated by surgical excision showed that the analysis based on 205 genes discriminated between patients who recurred distally and those who had no recurrence (Giri et al. 2006). While this was a small study involving only 15 patients, the findings clearly indicate that patients without the molecular signature associated with metastatic potential would be spared the toxicity of systemic therapy, thus pinpointing a role for genomic testing.

The sections below aim to discuss the main molecular targets and associated agents for targeted therapies in head and neck squamous cell carcinoma (HNSCC). Table 9.1 is a compilation of the currently trialled and/or pre-clinically tested agents designed for targeted therapies in HNC.

9.2 ANTI-EGFR TARGETED MOLECULAR THERAPIES

9.2.1 Biological Background

The epidermal growth factor receptor is a cell membrane (transmembrane) tyrosine kinase receptor that controls key cellular transduction pathways in healthy and malignant cells alike. EGFR plays an essential role in HNC development, progression, angiogenesis, and

TABLE 9.1   Molecular Targets and Associated Agents for Targeted Therapies in Head and Neck Squamous Cell Carcinoma (HNSCC)

Molecular Target	Agent	
EGFR	Monoclonal antibodies	Cetuximab
		Panitumumab
		Zalutumumab
		Nimotuzumab
	Tyrosine kinase inhibitors (small molecules)	Erlotinib
		Gefitinib
		Lapatinib
		Afatanib
		Dacomitinib
VEGF	Monoclonal antibodies	Bevacizumab
	Tyrosine kinase inhibitors (small molecules)	Sunitinib
		Sorafenib
		Vandetanib
Other targets/ pathways	Proteasomes inhibitors	Bortezomib
	PI3K/AKT/mTOR pathway inhibitors	Everolismus
		Temsirolismus
	Src kinase	Dasatinib
	Insulin-like growth factor receptor (IGF-1R)	
	Anti PD-1 (programmed cell death-1) monoclonal antibodies	Pembrolizumab
		Nivolumab

metastatic spread, through promotion of epidermal cell growth and regulation of cell proliferation (Mendelsohn & Baselga 2000). EGFR is part of a family of four growth factor receptors: EGFR (or ErbB-1) and the human epidermal receptors HER-2, HER-3, and HER-4 (Sedlacek 2000).

It is understood that most HNSCCs overexpress the epidermal growth factor receptor, which leads to increased tumour proliferation (Dassonville et al. 1993). Dassonville et al. (1993) determined the EGFR in tumour biopsies obtained from 109 consecutive HNC patients, while using control biopsies from 94 patients in a symmetric, non-tumoural area of the same anatomical location. The study concluded the following:

- All tumours had detectable levels of EGFR, with highly marked differences between patients
- In the majority of cases, the EGFR levels were higher in tumour samples than in biopsies originating from healthy individuals
- No significant difference in EGFR expression was found as a function of anatomic site or tumoural differentiation status
- A significant difference in EGFR distribution was observed between tumour stages I/II and III/IV

- No link between tumour EGFR levels and response to first-line chemotherapy by cisplatin and 5FU was found

- Overexpression of EGFR was correlated with shorter relapse-free and overall survival, making it an independent prognostic factor for poor clinical outcome.

These results were confirmed by further studies, showing that approximately 90% of HNSCCs exhibit overexpression of epidermal growth factor receptor (Kalyankrishna & Grandis 2006), a fact that might dictate the aggressiveness of these tumours. Additionally, Braunholz et al. (2016) demonstrated that EGFR regulates cell survival not only in the bulk tumour but also in the clusters of circulating tumour cells that can further lead to metastatic spread (Braunholz et al. 2016).

The two approaches implemented to date in order to restrain the functions of EGFR in cancer cells are: (1) the inactivation of monoclonal antibodies and (2) the inhibition of tyrosine kinase through small molecules (Figure 9.1). Anti-EGFR monoclonal antibodies prevent ligand binding by competitively linking to the extracellular area of EGFR, whereas small molecule inhibitors of the tyrosine kinase activity obstruct intracellular downstream signalling (Ciardello & Tortora 2001).

Pre-clinical studies also investigated the potential of EGFR blockade via mAbs to modulate tumour response to radiation (Harari & Huang 2001). Both in vitro and in vivo experiments showed that blockade of the EGFR system in squamous cell carcinoma lines can induce cell cycle arrest in the G_1 and G_2-M phases, while also reducing the S-phase fraction, the latter being known to be resistant to radiation. The accumulation of cells in the more radiosensitive phases of the cell cycle (G_1, G_2-M) enhances the overall radiosensitivity caused by monoclonal antibodies (the C225 mAb was used in this experiment). The modulation of radiation response by EGFR blockade was suggested to be due to several factors, such as: increased radiation-induced apoptosis, inhibition of DNA repair, growth arrest, and downregulation of angiogenic response (Huang & Harari 2000; Harari &

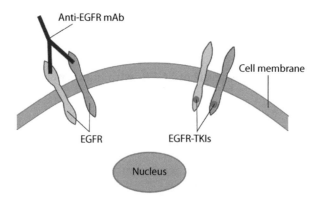

FIGURE 9.1   Representation on a cellular level of anti-EGFR therapy by monoclonal antibodies (mAb) and TKI.

Huang 2001). All of these features justify the clinical testing of anti-EGFR agents for HNC in combination with radiotherapy.

9.2.2 Clinical Aspects of Anti-EGFR Therapy

Since elevated levels of EGFR were shown to correlate with poor prognosis, this raised the need for the development of anti-EGFR agents. **Cetuximab** (see Table 9.1) is a monoclonal antibody designed to interact with EGFR and inhibit the function of the receptor. While widely trialled, the role of cetuximab in clinical settings is still contentious, due to varied clinical outcomes.

One of the first trials conducted with cetuximab in combination with radiotherapy for HNC patients showed improved outcome in terms of locoregional control and overall survival without increasing normal tissue toxicity compared with radiotherapy alone (Bonner et al. 2006). This landmark trial accrued 424 patients, with 213 patients being randomised to radiotherapy alone and 211 to radiotherapy combined with cetuximab. The results showed a notable difference with the addition of cetuximab to radiotherapy, as the anti-EGFR agent increased the 3-year locoregional control rates from 34% to 47% and the overall survival rates from 45% to 55%. Cetuximab-related normal tissue toxicity was observed as an acne-like rash. Connected to this adverse event, sub-group analysis revealed a rather interesting observation, whereby overall survival was improved in patients with a moderate to severe (Grade ≥2) rash compared with patients who had mild or no rash at all (Grade 1 or 0).

These findings have encouraged further clinical studies with cetuximab, by combining the agent with chemoradiotherapy. The Radiation Therapy Oncology Group (RTOG) 0522 randomised phase III trial reported by Ang et al. (2014) compared the addition of cetuximab with accelerated radiotherapy and cisplatin chemotherapy versus accelerated radiotherapy and cisplatin for locally advanced HNC. This large trial enrolled 891 patients with advanced disease (stage III to IV) with the aim of showing an improvement in outcome in the cetuximab arm. However, the results demonstrated no benefit from cetuximab, as no significant difference was found between the two clinical arms in terms of overall survival (72.9% vs 75.8% at 3 years), locoregional failure (19.9% vs 25.9%), or distant metastasis (13% vs 9.7%). Furthermore, the cetuximab arm required more frequent treatment interruptions (26.9% vs 15.1%). While p16-positive patients reported better outcome, the EGFR expression had no influence on the endpoint (Ang et al. 2014). A recent report of a sub-study conducted within the above trial indicated no clinical difference between the quality of life and performance status at baseline through to 5 years in the treated patients with or without cetuximab (Truong et al. 2017).

In alignment with the above results, an overview of phase III randomised clinical trials for HNC patients, showed that when added to radiotherapy, cetuximab offered improved locoregional control, whereas combined with platinum-based chemoradiotherapy, no further improvement has been observed (Specenier & Vermorken 2013). Given that cisplatin-based chemoradiotherapy is the standard of care in this patient group, the result is not supportive of the routine implementation of cetuximab.

To add more to the provocative data, a recent retrospective study undertaken in locally advanced HNC patients of Asian origins compared cisplatin-based radiotherapy with cetuximab-based radiotherapy (Rawat et al. 2017). Their results indicated that cisplatin is associated with superior disease-free survival (60% vs 14.3% at 3 years) and overall survival (74% vs 42.1% at 3 years) without significant differences in acute toxicities of all grades.

The combined effect of cisplatin and cetuximab was tested by the Eastern Cooperative Oncology Group, which analysed 117 patients with recurrent metastatic HNC randomly assigned to receive cisplatin with cetuximab or cisplatin with placebo (Burtness et al. 2005). The results showed a clear benefit from the addition of cetuximab to chemotherapy in terms of the objective response rate of 26.3% (cetuximab) vs 9.8% (placebo), however, neither progression-free survival nor overall survival showed any significant difference among the randomised groups. The potential effect of anti-EGFR therapy when combined with chemotherapy is confirmed by the results derived from the EXTREME (ErbituX in first-line Treatment of REcurrent or MEtastatic HNC) trial that investigated the HPV status and its impact on outcome in patients receiving chemotherapy with or without cetuximab (Erbitux) (Vermorken et al. 2014). It was shown that the addition of cetuximab to platinum-based chemotherapy improves survival in patients with recurrent and/or metastatic HNSCC regardless of the HPV status, although patients that tested positive for HPV had longer survival than non-HPV patients.

To understand the role of cetuximab in HPV+ patients, several phase III trials are undergoing that randomised HPV+ patients to receive either cisplatin or cetuximab with concurrent radiotherapy. Such trials are the RTOG 1016 (US), De-ESCALaTE HPV (UK) and TROG 12.01 (AUS) (still recruiting). The aim of these trials is to assess and compare the toxicities in the two arms and to evaluate whether substitution of cisplatin with cetuximab will result in similar locoregional control and overall survival. RTOG 1016 has enrolled over 800 patients with HPV+ oropharyngeal carcinoma, though more time is needed for outcome reports.

Panitumumab is another EGFR signal inhibitor, a fully human monoclonal antibody trialled in the clinics with varied success. In the CONCERT-1 phase II randomised trial of the Eastern Cooperative Oncology Group, 153 patients with locally advanced HNC were enrolled to receive cisplatin-based chemotherapy with or without panitumumab (Mesia et al. 2015). Superior locoregional control at 2 years was achieved in the group not receiving the anti-EGFR drug (68%) compared with the panitumumab/chemoradiotherapy group (61%), and also with worse side effects (40% vs 27% dysphagia, 55% vs 24% mucositis). The addition of panitumumab to cisplatin-based chemoradiotherapy was, therefore, unsuccessful in terms of outcome.

Panitumumab was also tested in patients with recurrent or metastatic HNC in combination with cisplatin and 5FU in an open-label, phase III, randomised trial (Vermorken et al. 2013) (see also Table 9.2). While overall survival did not improve among p16-positive patients, in p16-negative tumours, median overall survival was higher in the panitumumab group compared with chemotherapy alone (11.7 vs 8.6 months). A slight improvement in progression-free survival was also observed, although with greater toxicities. HPV status is suggested to be a predictive marker in patients treated with cisplatin+panitumumab.

TABLE 9.2   Randomised Phase III Clinical Trials Involving Anti-EGFR mAb Therapy for HNC

Trial Aim/Details	Anti-EGFR mAb Agent and Scheduling	Outcome/Conclusions
Eastern Cooperative Oncology Group phase III trial *for metastatic/ recurrent HNC patients* Cisplatin with/without cetuximab 121 patients (Burtness et al. 2005)	*Cetuximab* 200 mL/m^2 on day 1; subsequent cycles of 125 mL/m^2 each week. Cisplatin 100 mg/m^2 on day 1 every 4 weeks.	Improved outcome: **YES** Objective response in 26.3% cisplatin+cetuximab vs 9.8% cisplatin group.
Phase III trial *for advanced HNC patients* Radiotherapy (RT) with/without cetuximab 424 patients (Bonner 2006)	*Cetuximab* infusion one week before RT (400 mg/m^2) followed by weekly 250 mg/m^2 doses during RT course (70–72 Gy).	Improved outcome: **YES** Significant difference with the addition of cetuximab to radiotherapy—26% risk of death reduction; skin-related normal tissue toxicity
EXTREME phase III trial for *recurrent/metastatic HNC patients* Platinum-based therapy with/ without cetuximab 442 patients (Vermorken et al. 2008, 2014)	*Cetuximab* 400 mg/m^2 initial dose followed by 250 mg/m^2 per week for maximum 6 cycles; Cisplatin 100 mg/m^2 on day 1 or carboplatin + 5FU every 3 weeks for maximum 6 cycles.	Improved outcome: **YES** Median overall survival with cetuximab+chemotherapy (10.1 months) vs chemo-alone (7.4 months). Cetuximab increased response rate from 20% to 36%.
RTOG 0522 for *locally advanced HNC patients* Accelerated RT or IMRT + cisplatin with/without cetuximab 891 patients (Ang et al. 2014)	*Cetuximab* 400 mg/m^2 initial dose followed by 250 mg/m^2 per week during RT; Accelerated RT: 72 Gy in 42 fractions over 6 weeks, twice daily irradiation for 12 treatment days; IMRT: 70 Gy in 35 fractions over 6 weeks, twice daily doses once a week for 5 weeks; Cisplatin 100 mg/m^2 on days 1 and 22 of RT.	Improved outcome: **NO** No significant difference in overall survival (72.9% vs 75.8% at 3 years), locoregional failure (19.9% vs 25.9%), or distant metastasis (13% vs 9.7%) **Obs:** HPV+ patients had better outcome irrespective of EGFR status.
SPECTRUM phase III trial for *recurrent/metastatic HNC patients* Platinum-based therapy with/ without panitumumab 657 patients (Vermorken et al. 2013)	*Panitumumab* 9 mg/kg on day 1 of each cycle; Cisplatin 100 mg/m^2 on day 1 of each cycle, up to six 3-week cycles; 5FU 1000 mg/m^2 on days 1–4 of each cycle.	Improved outcome: **SMALL** Progression-free survival with panitumumab was 5.8 months vs 4.6 months in the control group. Severe toxicities were more frequent in the panitumumab group.
Canadian Cancer Trials Group study HN.6 phase III trial for *locoregionally advanced HNC patients* Cisplatin-based standard RT vs accelerated RT + panitumumab 320 patients (Siu et al. 2016)	*Panitumumab* 9 mg/kg for 3 doses; Standard RT: 70 Gy, 2 Gy/day over 7 weeks; Accelerated RT: 70 Gy, 2 Gy/day, 6 weeks; Cisplatin 100 mg/m^2 for 3 doses.	Improved outcome: **NO** 2-year overall survival cisplatin-RT (85%) vs panitumumab-RT (88%); progression-free survival 73% vs 76%; grade 3–5 adverse events 88% vs 92%.
Phase III trial for *incurable (with standard therapy) HNC patients* Best supportive care with/without zalutumumab 286 patients (Machiels et al. 2011)	*Zalutumumab* initial dose of 8 mg/ kg + 2 weekly doses of 4 mg/kg. Patients with 0/grade 1 rash received + 4 mg/kg every 2 weeks up to 16 mg/kg. Control group: methotrexate up to 50 mg/m^2 per week.	Improved outcome: **SMALL** Median overall survival with zalutumumab (6.7 months) vs control group (5.2 months); progression-free survival 9.9 vs 8.4 weeks.

No improvement in clinical outcome with panitumumab was found by the Canadian Cancer Trials Group study HN.6, which randomised patients to receive either standard cisplatin-based chemoradiotherapy or accelerated radiotherapy with panitumumab in locoregionally advanced HNC (Siu et al. 2016). The 2-year, progression-free survival in the anti-EGFR arm was 76% compared with 73% in the cisplatin-radiotherapy arm. Toxicity was also higher in the panitumumab group (92% vs 88%).

Zalutumumab is another human mAb targeting EGFR, which was investigated in a phase III, randomised trial with the aim of improving outcome in recurrent/metastatic HNC patients who failed to respond to platinum-based chemotherapy (Machiels et al. 2011). While no difference in overall survival was seen, progression-free survival was increased in the zalutumumab group (Table 9.2).

A recent study involving **nimotuzumab** in the management of unresectable advanced HNSCC patients, provided survival benefit when concurrently given with chemoradiotherapy compared with chemoradiotherapy alone (overall survival at 60 months 57% vs 26%) (Reddy et al. 2014).

An Asian phase III clinical trial is currently recruiting patients to evaluate the therapeutic advantage of nimotuzumab as an adjuvant treatment to standard chemoradiotherapy in resectable HNC patients (NCT00957086).

The **small molecules** tested for anti-EGFR therapy (Table 9.1) are mostly investigational, as they are not approved for widespread use in HNC management. Nevertheless, many are under exploration in clinical trials, and several of them reached phase III levels (Table 9.3).

As shown in Table 9.3, most EGFR-TKIs are tested for metastatic or recurrent patients, either as an adjuvant therapy or as a replacement for a standard chemotherapeutic agent in order to improve treatment outcome. While the results reported by trials often show only marginal improvement with the addition of the currently available EGFR-TKI small molecules, the toxicity profile of these drugs is acceptable and justifies more research in this direction (Yunhong et al. 2017).

One way to increase the efficacy of targeted therapies is to take full advantage of next-generation sequencing technology to perfect patient selection by identifying an EGFR sensitivity signature (Psyrri & Dafni 2014). Furthering the idea of patient stratification, a recent study recommends adequate selection for cetuximab-based therapy given that its pharmacokinetics influences overall survival in HNC patients (Pointreau et al. 2016). Regarding adverse events, the most common toxicity caused by EGFR inhibitors is dermatological, although the mechanism behind this is not fully elucidated. Therefore, more research is warranted with EGFR inhibitors to better manage both tumour control and normal tissue toxicities.

9.3 ANTI-ANGIOGENIC TARGETED THERAPIES

9.3.1 Biological Background

Angiogenesis, the formation of new blood vessels, is a critical process in the progression of solid tumours (Hicklin & Ellis 2005). Consequently, attempts have been made to target this process to hamper the development of tumours leading to the use of anti-angiogenic targeted therapies.

TABLE 9.3   Randomised Phase II/III Clinical Trials Involving Anti-EGFR TKI Therapy for HNC

Trial Aim/Details	Anti-EGFR TKI Agent and Scheduling	Outcome/Conclusions
Eastern Cooperative Oncology Group phase III trial *for metastatic/recurrent HNC patients* Docetaxel with/without gefitinib 239 patients (Argiris et al. 2013)	*Gefitinib* 250 mg daily tablet until disease progression Docetaxel 35 mg/m² weekly, on days 1, 8, and 15 of a 28-day cycle Control group: docetaxel weekly dose + placebo.	Improved outcome: **NO** Median overall survival in the gefitinib group 7.3 months vs 6 months in the control ($p = 0.6$). Median time-to-progression 3.5 months vs 2.1 months ($p = 0.19$). Younger patients (<65 years) had a higher survival benefit with gefitinib than older ones.
Phase II randomised trial *for locally advanced HNC patients* Cisplatin-radiotherapy with/ without erlotinib 204 patients (Martins et al. 2013)	*Erlotinib* 150 mg daily, starting 1 week pre-RT Standard radiotherapy: 70 Gy, 2 Gy/day Cisplatin 100 mg/m² on days 1, 22 and 43 of RT.	Improved outcome: **SLIGHT** Complete response rate in erlotinib vs control arm 52% vs 40% ($p = 0.8$). No difference in progression-free survival after 26 months follow-up. No increase in high grade toxicity.
Phase III trial *for high-risk HNC patients* Adjuvant post-operative lapatibin with concurrent chemoradiotherapy (CRT) 688 patients (Harrington et al. 2015)	*Lapatinib* 1500 mg daily up to 1 week prior CRT + during CRT + 12 months maintenance monotherapy Radiotherapy 66 Gy, 2 Gy/day, 5 days/week Cisplatin 100 mg/m² on days 1, 22 and 43 of RT.	Improved outcome: **NO** The study ended early with a median follow-up time of 35.3 months due to plateauing of disease-free survival events (DFS). No significant difference in DFS—irrespective of HPV status. Serious adverse events: 48% (lapatinib) vs 40% (control).
LUX-Head & Neck 1 phase III trial *for metastatic/recurrent HNC patients* Afatinib vs methotrexate 483 patients (Machiels et al. 2015)	*Afatinib* 40 mg daily Methotrexate 40 mg/m² per week intravenous infusion.	Improved outcome: **YES** Progression-free survival in the afatinib vs control arms after 6.7 months median follow-up: 2.6 vs 1.7 months ($p = 0.03$). Serious adverse events: 14% (afatinib) vs 11% (control).

Angiogenesis is intrinsically dependent on the development of hypoxia in the tumour microenvironment, when a cascade of proangiogenic molecules is released determining the formation of neovasculature. Vascular endothelial growth factor A (VEGF-A) is the most known of these molecules. It is a vascular permeability factor belonging to the platelet derived growth factor (PDGF) superfamily which also includes VEGF-B, VEGF-C, VEGF-D and placental growth factor (PIGF) (Ferrara 2005; Christopoulos et al. 2011). The expression of VEGF has been reported in many solid tumours, including squamous cell carcinomas of the head and neck. The overexpression of this molecule has been associated with increased tumour progression, increased resistance to chemotherapy, positive lymph node metastasis and poor prognosis (Riedel et al. 2000; Smith et al. 2000; Tse et al. 2007; Boonkitticharoen et al. 2008). The key mediator modulating VEGF expression in solid tumours is hypoxia-inducible factor (HIF-1α). The VEGF family signals through a group of cell surface receptor tyrosine kinases that include VEGFR-1, VEGFR-2 and VEGFR-3. Of these, VEGFR-2 has been identified as the principal angiogenic receptor. VEGF also interacts with neuropilins, a family of coreceptors that include NRP-1 and NRP-2. The neuropilins are thought to form complexes with the VEGF receptors, thus enhancing their biological activities (Ferrara, Gerber & LeCouter 2003; Christopoulos et al. 2011).

From the perspective of the signalling pathway of VEGF, three possible inhibition mechanisms have been proposed and investigated. Thus, the VEGF-A ligand can be sequestered using a monoclonal antibody against VEGF, the ligand-receptor interaction can be inhibited with a monoclonal antibody against VEGFR or the kinase activity of the receptor can be blocked with a TKI against the adenosine triphosphate (ATP)-binding site of the receptor (Christopoulos et al. 2011). Specific inhibitors for each of these approaches have already been approved for treatment or are under investigation in clinical trials. Thus, Bevacizumab, also known as Avastin, a recombinant humanised monoclonal antibody, has been used for the first mode of inactivation against VEGF, while Imclone-1121b is a monoclonal antibody used against VEGFR-2. With respect to TKI, these are selective rather than specific to those of the VEGFR and this has been associated with increased off-target effects.

In this context, it is worth mentioning that even though anti-angiogenic therapies have shown promising results in decreasing tumour vessel growth in preclinical studies (Kim et al. 1993; Riedel et al. 2003), when administered as monotherapy in clinical studies, they have yet not shown long-term survival benefits. It is in combination therapies that the anti-angiogenic targeted therapies have shown potential for improvement (Christopoulos et al. 2011). The enhancement of chemoradiation therapy effects may seem paradoxical at first, since hampering the formation of neovasculature in tumours will diminish the drug delivery to the cancer cells and will also increase tumour resistance to radiation by promoting hypoxia. Nevertheless, it has been postulated that the administration of antiangiogenic agents would normalise tumour vasculature, e.g. by increasing pericyte coverage, which would thus become less leaky and chaotic (Naumov et al. 2009). This will in turn increase the delivery of oxygen and cytotoxic drugs to the cells. Another hypothesis states that inhibition of the protein kinase signalling pathways would increase cytotoxic effects by lowering the pro-apoptotic threshold of tumour cells (Epstein 2007).

Another important aspect in quantifying the efficacy of anti-angiogenic targeted therapies is that the molecular mechanisms by which tumours induce angiogenesis may be different (Codeca et al. 2012). Thus, it has been shown that the expression patterns for cytokines mediating angiogenesis are different in HNC patients (Hasina et al. 2008). This, in turn, suggests that the inhibition of a specific molecular pathway can interfere with the angiogenesis process and consequently increase the effectiveness of the therapeutic approach, only if the target of the therapy is expressed by the tumour cells.

9.3.2 Clinical Studies of Anti-Angiogenic Targeted Therapies

Bevacizumab is the most extensively studied anti-angiogenic agent, however, mostly in phase I or II studies. Nevertheless, it has been shown that radiotherapy with concurrent weekly docetaxel and biweekly bevacizumab is tolerable and effective in HNSCC.

Employing 30 patients with previously untreated locally advanced HNSCC, Yao et al. (2015) showed that the combined therapy led to the 61.7% progression free survival, 68.2% overall survival, 84.5% locoregional recurrence-free survival and 80.5% distant metastasis-free survival for a median follow-up of 38 months. They also reported mucositis and dermatitis as local toxicities, with two patients developing haemorrhage.

Bevacizumab has also made the object of a large phase III study investigating its effectiveness in combination with chemotherapy (Argiris et al. 2017). Investigating the response in 403 patients in two arms with a median follow-up of 23.1 month, the authors reported that the addition of bevacizumab increased the median overall survival from 11 months to 12.6 months and the progression free survival from 4.4 months to 6.1 months. However, the rate of grade 3–5 bleeding was doubled, 7.7% in the bevacizumab arm versus 3.5% in the non-bevacizumab arm.

Other studies were dedicated to combined therapy approaches. Thus, Cohen et al. (2009) reported on the combination of bevacizumab and erlotinib, a small molecule inhibitor of the EGFR tyrosine kinase, in patients with recurrent or metastatic HNSCC. For a population of 48 patients, they have found a median time of overall survival of 7.1 months and a median time of progression-free survival of 4.1 months. Similar results were also found by combining bevacizumab and cetuximab, an IgG1 monoclonal antibody against EGFR (Gibson et al. 2009). Another study investigated the combination of bevacizumab and cisplatin and reported that in a population of 49 patients, the locoregional control rate for a median follow-up of 9 months was 100%, with estimated one-year progression free survival of 83% and estimated one-year overall survival of 88% (Pfister et al. 2009).

In contrast, studies investigating the efficacy of antiangiogenic TKI have failed to show an improvement. Thus, the administration of sunitinib and sorafenib as single agents in patients with recurrent or metastatic disease was shown to be well tolerated, but the response rates were rather poor (Elser et al. 2007; Fountzilas 2009; Choong et al. 2010; Machiels et al. 2010; Williamson et al. 2010).

9.4 OTHER TARGETS AND PATHWAYS

So far, inhibition of the EGF receptor with cetuximab is the only targeted biological agent approved for use in HNC and, as has been described earlier in this chapter. Cetuximab is now a drug included in the therapeutic armamentarium together with radiotherapy for locally advanced, mostly non-resectable tumours and also in the systemic treatment together with cisplatin and 5FU for recurrent or distantly metastatic tumours. However, multiple pathways have been identified that are deregulated in HNC patients and therefore could be candidates for genotype matching and interaction with radiotherapy. Results from ongoing HNC sequencing analyses, including data from The Cancer Genome Atlas initiative will provide a platform for rational experimental and subsequent prospective, clinical studies (Cancer Genome Atlas Network 2015).

9.4.1 Proteasomes Inhibitors

The proteasome pathway is the major nonlysosomal proteolytic system in the cell's cytosol and nucleus, triggering degradation of most proteins involved in several mechanisms, such as apoptosis, inflammation, cell cycle regulation and immunosurveillance (Roccaro et al. 2006). Inhibition of this pathway leads to apoptosis by disrupting signalling pathways induced by the presence of NF-kappa-B transcription factor C.

Bortezomib is a small proteasome inhibitor targeting the 26S proteasome, a large proteasome complex that blocks the activation of NF-kappa-B (Adams & Kauffman 2004). Bortezomib also induces phosphatase and tensin homolog (PTEN) expression, leading to down regulation of PI3K/Akt signalling and therefore also enables radiosensitisation (Fujita et al. 2006). The hypothesis of radiosensitisation was tested in a system of radioresistant HNC cell lines and it was found that bortezomib could induce radiosensitisation and did so in conjunction with decreased pAkt (Weber et al. 2007). Bortezomib also induces an ultimate enhancement of apoptosis as evidenced by the accumulation of cleaved poly ADP ribose polymerase (PARP) and also inhibits angiogenesis (Tamatani et al. 2013). It is the first agent of its kind to be selected for clinical use and is FDA-approved for the treatment of multiple myeloma and mantle cell lymphoma. The agent has shown anti-tumour activity for several tumour types, including HNSCC. Bortezomib blocks NF-kappa-B activation and proliferation in HNC cell lines, induces apoptosis, inhibits growth and angiogenesis and, as mentioned above, also circumvents at least partial radioresistance in syngeneic murine squamous carcinoma cells and human squamous cell carcinoma xenograft models (Lun et al. 2005).

Bortezomib has therefore been tested in a feasibility study together with radiation in patients with recurrent HNC. Radiotherapy was given in a radical dose of 60 Gy, fractionated 5 days per week and bortezomib was given in two dose levels; nine patients were enrolled in the study, all of whom had previous surgery and radiotherapy. Three patients had transient tumour regression and another three had stable disease for 3 or more months. In this study, the authors could decide a maximum tolerated dose (MTD) with a grade 3 toxicity of hypotension and hyponatremia (van Waes et al. 2005).

Several following phase I studies have tried to evaluate the use of bortezomib alone or in combination with other agents, chemotherapeutic drugs or radiation, for patients with HNC. However, most studies were performed on advanced solid tumours, most often with few patients with HNC. To further explore the role of bortezomib in treatment of advanced and recurrent HNC a phase I dose-escalation study using bortezomib in combination with chemoradiotherapy including cisplatin was performed (Kubicek et al. 2012). Twenty-seven patients were enrolled in the study, 10 with locally advanced and 17 with recurrent tumours; 9 with previous chemoradiation and 8 with previous radiation. Despite the small number of patients in the study, results were considered promising with recommendations of bortezomib doses for future trials. There was a median survival of 15.5 months in patients with previous radiotherapy and 48.4 months in radiation naive-patients. In this study of chemoradiotherapy with cisplatin together with bortezomib, a maximal tolerated dose could be established and the dose-limiting toxicity was hematologic.

In conclusion, inhibition of this pathway with bortezomib leads to radiosensitisation via upregulation of PTEN activity, the subsequent down-regulation of phosphorylated AKT, enhancement of apoptosis, as evidenced by the accumulation of cleaved PARP, and an inhibition of angiogenesis. Evidence to support the introduction of these agents together with radiotherapy in the clinic is still lacking.

9.4.2 PI3K/AKT/mTOR Pathway Inhibitors

The PI3K/AKT/mTOR pathway plays a major role in different cell processes, including protein synthesis and cell survival. It is a signalling cascade downstream of the EGFR that stimulates cell growth and is often dysregulated in tumour cells. Targeting this pathway is, therefore, one potential method for overcoming resistance to anti-EGFR targeted therapy. The pathway can be activated by the upstream activation of tyrosine kinase receptors such as EGFR and IGF-1R. mTor is a serine/threonine protein kinase that is involved in regulation of cell growth, proliferation of cells, cell motility and protein synthesis and has been shown to be activated in up to 80% of patients with HNC (Williamson et al. 2010). Genetic aberrations of the PI3K pathway are common in HNC cells. One of the isoforms of the 110 kDa catalytic subunit is encoded by the PIK3CA gene, which is mutated in up to 20% of patients with HNC, especially through gene amplification and low-level copy number increase. It has been found to be particularly common in HPV+ patients (Stransky et al. 2011).

Also, PTEN mutations have been reported in 7% of HNC. The mTOR pathway is frequently activated, independent from activation of EGFR or the presence of mutant p53, particularly in HPV+ tumours (Stransky et al. 2011).

Temsirolismus has been found to radiosensitise head and neck squamous cancer cells in both in vitro and in vivo model systems (Ekshyyan et al. 2010). The combination of this drug and radiation was even more effective in tumour growth inhibition than cisplatin and radiation in murine xenograft models. The addition of temsirolimus to radiation increased apoptosis, decreased the number of tumour blood vessels and reduced radiation-induced activation of the mTOR pathway. By using a dual inhibitor of the pathway in a variety of cancer models in vitro and in xenografts, further knowledge regarding mechanisms behind radiosensitisation could be identified: the process of DNA damage repair could be attenuated by decreasing radiation-induced phosphorylation of DNA-PKcs and ATM and by the promotion of autophagy of the tumour cells.

A phase I study was initiated to determine the maximum tolerated dose, MTD, of daily **everolismus** combined with weekly cisplatin and radiotherapy in patients with locoregionally advanced HNC. Common adverse events were mucositis, pain in oral cavity or pharynx, dysphagia and hyperglycemia; dose-limiting toxicity was in this study lymphopenia (Fury et al. 2013).

Inhibitors of the PI3K pathway are also being studied in ongoing phase II trials in HNC together with cetuximab and chemotherapy. AKT inhibitors are being studied in recurrent and metastatic cancer of the nasopharynx and mTor inhibitors such as rapamycin, everolimus and temsirolimus are being assessed for use in patients with HNC in phase II studies in neoadjuvant and recurrent/metastatic settings.

In conclusion, inhibitors of this pathway radiosensitises tumour cells preferably by directly decreasing DNA-PKcs and mutated ATM to inhibit DNA damage repair and increase apoptosis; inhibitors of mTOR also attenuate tumour angiogenesis and seem to induce increased cell death through processes related to autophagia.

9.4.3 Src Family Kinases

The role of Src family kinases in cell signalling and morphologic transformation processes is well established. Nine members of the Src kinase family have been identified and they all belong to a family of non-receptor kinases that are involved in a number of signalling pathways controlling multiple functions such as cell adhesion, cell mobility, cell proliferation and terminal differentiation. They directly regulate several important signalling pathways including Ras/MAPK, PI3K/Akt and STAT3 (Martin 2001) and also play an important role in VEGF induced angiogenesis. The Src kinases are tyrosine kinases and participate in signal transduction in response to several stimuli, including angiogenic factors. The first defined proto-oncogene, c-Src is often overexpressed in many human malignant tumours, including HNC. Expression of activated c-Src is higher in tissues of human HNC than in normal mucosa and correlates with an invasive, poorly differentiated phenotype with more advanced nodal stage (Mandal et al. 2008).

Several of identified members of the Src family are also expressed and activated by TGF-alfa in HNC cell lines (Bu et al. 1996). When c-Src is overexpressed, the tumours were found to grow more rapidly and have a higher propensity for invasion and metastatic potential (Myoui et al. 2003).

To target molecules or receptors known to interact with EGFR could further improve treatment outcome. Such a molecule is represented by c-Src kinase. Therefore, one study investigated **dasatinib**, an inhibitor of c-Src kinases for its efficacy to enhance radiosensitivity of human HNSCC in vitro, also trying to identify potential underlying mechanisms for such an effect. The results showed that dasatinib induced apoptosis and blocked DNA repair in EGFR-expressing cells and improved radiotherapy outcome (Raju, Riesterer & Wang 2012).

Due to the promising preclinical evidence for the combination of dasatinib with radiation, further studies of this combination in clinical trials have been warranted. An ongoing phase II study has been designed to define MTD and tumour response of daily oral dasatinib in combination with cetuximab and radiotherapy in stage II and III HNC patients or with cisplatin/cetuximab and radiotherapy in stage III and IV patients. The results of these studies have not yet been published.

9.4.4 Insulin-Like Growth Factor Receptor, IGF-1R

The receptor for the IGF-1R belongs to the insulin receptor subfamily of RTKs. IGF-1R is a teramorphic transmembrane receptor tyrosine kinase that is widely expressed in human tissue (Adams et al. 2000). Binding of endogenous ligands such as IGF-1R or IGF-II initiates the activation of downstream signalling cascades including PI3K/AKT/mTOR pathway and MAPK. IGF-1R mediates a variety of important cellular events such as cell proliferation, differentiation, motility, resistance to apoptosis and neovascularisation (Barnes et al. 2007). It has been reported that IGF-1R is overexpressed in HNC cell lines and that IGF-1R signalling was associated with the tumourigenicity, proliferation and motility of these cell lines (Barnes et al. 2007). Other investigators found that IGF-1R was overexpressed in 73% in tumours from patients with HNC and was correlated with poor survival in patients with advanced disease. This finding suggested that IGF-1R could be one relevant target to try to inhibit when treating HNC (Jun et al. 2009).

Promising preclinical results were also achieved with an anti-IGF-1R human antibody in HNC cell lines and tumour xenografts (Barnes et al. 2007). However, two trials, one for patients with progressive disease on platinum-based therapy and one for patients with recurrent or metastatic disease, were negative with respect to response and survival when patients were given similar human monoclonal antibody targeting the IGF-1R. Despite promising early trials in other malignancies such as non-small cell lung cancer, Ewing's sarcoma, colorectal, breast and prostate cancer and the preclinical studies revealing that IGF-1R is overexpressed in HNC no positive clinical results have been described so far, either by targeting the receptor itself or in combination with radiotherapy. Interaction between EGFR and IGF-1R has been established in many cancer types. To determine if co-targeting these receptor pathways could overcome cellular resistance to radiation, cetuximab and IMC-A12, an inhibitor of IGF-1R, were combined when irradiating HNC cell lines. Concurrent treatment regimen failed to further enhance tumour response to cetuximab and/or radiation. There was no effect on cell survival by the inhibitor of IGF-1R but it increased radiosensitivity in one of the cell lines (Raju et al. 2015).

In conclusion, no clinical data have been presented so far to describe a positive gain with respect to tumour response or survival when inhibition of IGF-1R was combined with radiotherapy for patients with HNC.

9.4.5 PD-1 (Programmed Cell Death-1) Receptor

The PD-1 receptor together with its two ligands (PD-L1 and PD-L2) play a significant role in T-cell regulation to such extent that the inhibition of the PD-1 protein can activate T cells that trigger immune responses against malignant cells. Of particular importance is the PD-1/PD-L1 pathway that regulates T cell activity during inflammatory response to infection. PD-L1 is expressed by several cancers, including squamous cell carcinomas of the head and neck. This information is important since cancer cells that express PD-L1 were shown to evade the host immune system, while inhibition of PD-1 enhances tumour response to treatment (Müller et al. 2017).

Furthermore, PD-L1 expression was demonstrated to have a prognostic value in HNC, irrespective of the anatomical localisation. Müller et al. showed that patients with a high PD-L1 expression and activation of the PD-1/PD-L1 pathway exhibit a highly aggressive phenotype of HNC, independent of tumour stage, grade or even size (Müller et al. 2017).

Drugs that target this pathway are already under clinical trials, with highly promising early results. The two anti-PD-1 monoclonal antibodies trialled in patients with recurrent or metastatic HNC were nivolumab and pembrolizumab. In a phase III clinical trial with **nivolumab** versus standard single-agent therapy, nivolumab doubled the 1-year survival among platinum-refractory cancer patients (Harrington et al. 2017). **Pembrolizumab** was studied within an open-label, multicentre, phase 1b trial with the aim to assess safety, tolerability and antitumour activity (Seiwert et al. 2016). Pembrolizumab was administered to patients that tested positive for the PD-L1 ligand (60 patients). HPV+ patients showed better outcome, as overall response to the PD-L1 inhibitor was 25% among HPV+ patients and 18% among the whole group. Being well tolerated, pembrolizumab supports further clinical studies in advanced HNC patients.

9.5 CONCLUSIONS AND FUTURE PROSPECTS

To overcome the limitations of conventional chemo-radiotherapy concerning long-term tumour control as well as adverse events, molecular targeted therapies have been developed and tested for HNC. Resulting from the above clinical trials, the tested drugs showed limited success. Reasons for this limitation are inadequate patient selection and resistance to treatment.

The identification of individual tumour characteristics using specific biomarkers should be a key attribute when selecting patients for targeted therapy. Otherwise, resistance to treatment can occur that hinders optimal response. While the mechanisms behind resistance to molecular targeted therapy are complex and far from being fully elucidated, certain mechanisms have been identified for anti-EGFR therapy and they include: activation or upregulation of ErbB family members other than EGFR, phenotypic transformation, activation of parallel downstream signalling pathways, induction of epithelial to mesenchymal transitions, and resistance to apoptotic cell death (Campbell et al. 2016; Yamaoka, Ohba & Ohmori 2017).

Similarly, resistance to VEGF receptor inhibitors is the greatest challenge among patients treated with anti-angiogenic agents. Even when initially successful, the long-term results of anti-VEGFR therapy are poor, as often cancers escape from anti-angiogenic therapy and they relapse. Among the mechanisms behind treatment resistance the following can be mentioned: treatment-induced hypoxia that mediates resistance at the interface between tumour and host, amplification of pro-angiogenic genes by the targeted cancer cells and secretion of pro-angiogenic factors, escape through different vascularisation modes such as vasculogenesis, vascular mimicry and vessel co-option (Loges, Schmidt & Carmeliet 2010).

A solution to the challenges imposed by treatment resistance is the identification of new alternative targets and/or the co-administration of agents that display different mechanisms of action. For instance, cetuximab-resistant HNC exhibited substantial tumour response when co-targeting with mTOR inhibitors by affecting downstream pathways that remain activated during EGFR blockade (Wang et al. 2017). When EGFR resistance is driven by the upregulation of other ErbB family members, inhibition/disruption of ErbB-2/ErbB-3 heterodimerisation was shown to restore cetuximab sensitivity both in vitro and in vivo (Yonesaka et al. 2011). Co-targeting of EGFR and hedgehog pathways in HNC was able to reduce colony formation and cellular proliferation (Liebig et al. 2017).

Regarding alternative targets for angiogenesis inhibition, certain strategies are currently under investigation such as: targeting *Notch* regulatory pathways, induction of vessel normalisation through PHD2 (an oxygen sensor within endothelial cells) inhibition leading to improved oxygenation, inhibition of vessel maturation by blocking of neuropilins (molecules that regulate angiogenesis as co-receptors for VEGF) (Loges, Schmidt & Carmeliet 2010).

In conclusion, adequate patient selection using well-targeted biomarkers is the key to successful outcome in targeted therapies.

CHAPTER 10

Retreatment Issues

10.1 CAUSES OF TUMOUR RECURRENCE

Malignant tumours in the head and neck area belong to those malignancies where treatment results have continually improved. Over the last several years, treatment in most institutions consisted of surgery for resectable tumours with postoperative radiotherapy, according to special indications and radiotherapy only for advanced, unresectable disease. This combined treatment, however, often implied that important organs or tissues had to be removed, resulting in major loss of function. Especially during the two last decades, treatment regimens (schedules) have been established with radiotherapy as the cornerstone therapy, given in combination with systemic agents, classical cytostatics (cisplatin), or targeted drugs (cetuximab) in order to maintain important functions, mainly salivation, swallowing, and chewing but also preserving cosmesis.

Such new regimens have been successful and have constituted a paradigm shift for the care of patients with these tumours, especially for those with locoregionally advanced disease. Several meta-analyses have indicated an overall survival advantage at 5 years of >8% with such strategies. The reason for the better survival is the improved tumour control based on an implementation of important radiobiological mechanisms, proved relevant in prospective, randomised studies, such as altered radiation fractionation of the radiation dose and in the basic knowledge of how ionising radiation interacts with different kinds of systemic agents. However, one of the most important reasons for this treatment success can be found in the development of practical radiotherapy delivery, with three-dimensional conformal radiation therapy (3D-CRT) and later, more conformal techniques, represented by IMRT, stereotactic body radiation therapy (SBRT), modern brachytherapy, and intensity modulated proton therapy (IMPT). The possibility of using these newer techniques and better targeting tissue has gained from the mutual development in diagnostic radiology, improved computed tomography (CT), magnetic resonance imaging MRI and positron emission tomography (PET), enabling a more precise delineation of malignant tissue from normal structures.

Despite a number of advances after such primary, curative-intent radiotherapy regimens over the last decades concerning both tumour control and organ preservation, ~30%–50% of patients with locally advanced squamous cell carcinoma develop locoregional recurrence and most often are within the high-dose treatment volume. Moreover, patients who are cured of their tumour have an increased risk of developing a second metachronous tumour in the head and neck area, which occurs in up to 40% of definitely cured patients, sometimes or most often, in close proximity to the primary tumour. From a recent meta-analysis of radiotherapy combined with chemotherapy in head and neck cancer, comprising data from 50 trials of concomitant chemoradiotherapy and 30 trials with induction chemotherapy respectively, the rates of local recurrence at 5 years, with or without regional recurrence, was 50.8% and 47.5% in the experimental arms and 60.1% and 46.5% in the control arms, where patients had radiotherapy only (Blanchard et al. 2011).

Classically, surgery is proposed as the treatment of choice for patients presenting with a locoregional recurrence or a second primary. Surgical salvage should be considered the most effective treatment when cure is intended and is the treatment of choice for patients who are operable and whose tumours are resectable. For surgically treated patients with recurrent cancer of the tonsil, between 25% and 45% of patients have been described as long-term, disease-free survivors (Bachar et al. 2010) and according to a meta-analysis with a total of 1,080 patients, a survival rate of 39% could be expected at 5 years after salvage surgery. Although the authors present these results as encouraging, they cautioned that the "costs were great" with respect to the complications, infections, bleeding, and wound healing problems (Goodwin 2000).

The best chance for a cure was reported for patients with early-stage recurrent tumours, whereas those with T3/T4-reclassified disease should be considered poor candidates to undergo salvage surgery. In this meta-analysis, the impact of treated site was less important than recurrence stage. Therefore, in reality, few patients are candidates for curative resection. Considering the general health status of patients, morbidity after previous therapies, preference of patients and extent of recurrent disease, most studies describe less than half of all patients amenable to radical salvage treatment (Taussky, Dulguerov & Allal 2005; Nichols et al. 2011).

It is also a widespread experience that even if patients present with seemingly resectable lesions, surgery can be technically most challenging because of difficulties to operate in previously irradiated or otherwise manipulated tissue and the proximity of tumour to critical structures such as the carotid artery or skull base. Complete resections are, therefore, difficult to achieve for many patients.

10.2 APPROPRIATENESS OF REIRRADIATION

When patients present with the most common situations of (a) unresectable disease or are otherwise unsuitable for surgery or (b) have had incomplete resections after surgery attempted to be radical, three options are discussed: supportive care, chemotherapy alone, or radiotherapy with or without systemic therapy. Several studies have shown that if the patient is left untreated, the prognosis is poor, with a median survival of only a few months (Kotwall et al. 1997). Locoregional recurrences can also affect the patients' quality of life as a morbid reminder of disease presence apart from being a source of functional impairment that affects swallowing and phonation. Patients with locoregionally uncontrolled head and neck tumours often have an unpleasant end-of-life situation. They usually present with symptoms that are difficult to decrease, especially pain and extreme periods of bleeding from the tumour mass. The fact that the tumour in this region can be visible and that there may exist an odour of the often-necrotic and infected tumour mass can have a profound negative effect on the patient's self-esteem.

When chemotherapy is chosen as a single therapeutic modality, the therapeutic intention is palliative. Many drugs, such as methotrexate, platinum analogues, and cetuximab can shrink the tumour tissue, although not in >10%–40% of cases, and can, therefore, have palliative effects and temporarily reduce pain and improve functions as described above. However, it is known from randomised trials including both recurrent and primarily metastasised tumours that long-term survival at 5 years is usually <4% and median survival only 5–7 months. A prospective randomised phase III study for the same group of patients showed an improvement in median survival with the addition of cetuximab to the common combination of platinum and 5-fluorouracil, but still with few patients alive at 5 years (Vermorken et al. 2008). Thus, the role of reirradiation both with respect to improved survival and quality of life could be considered.

Evidently, there have been major concerns with reirradiation principally regarding two issues:

1. How can a second series of radiotherapy to the same region be safely performed with similarly high radiation doses since the tolerance doses of the normal tissues should already have been administered, i.e., normal tissue constraints had been passed?
2. In the situation in which reirradiation is considered for a true local recurrence (and not for a second primary in the vicinity of the first tumour), how can the same or similar regimen be expected to have a curative potential? And if so, what could the mechanisms be?

For the patient presenting with a recurrent tumour and evaluated for a reirradiation protocol, both early and late tissue reactions from the new treatment must be considered. The decision of reirradiation must be based on risk factors relating to such reactions, potentially both threatening survival (acute and late effects respectively) and producing severe late tissue complications with a severe negative impact on patient's quality of life. Preclinical data on reirradiation effects in the upper airway mucosa are lacking. However, mucositis in the oral cavity and oropharynx has been quantified in patients after repeated courses of radiotherapy with different treatment intermissions. If treatment intermissions are at least 2 weeks long, the mucositis appears with an identical time course and severity after each of three treatment courses (Van der Schueren et al. 1990). When the breaks in treatment are shorter, the severity of mucositis can be lower during the following course, probably because of increased repopulation counteracting the cell kill from the new irradiation course (Maciejewski et al. 1991). More severe reactions, grades 3 and 4, are observed earlier after reirradiation, probably reflecting an increased mucosal vulnerability with atrophy, cell depletion, reduction in the amount of vessels, etc. Late tissue reactions are partly reflected in how well stem cells in previously irradiated tissue volume survive or will migrate into the volume from adjacent non-irradiated sites and will define a partial or complete restoration of subsequent radiation tolerance.

However, it has become increasingly evident with time that reirradiation protocols, with or without concomitant systemic chemotherapy, can result in durable disease control, both when administered after surgery or as sole treatment. In the era of conventional radiation techniques, before the introduction of IMRT or SBRT, several retrospective studies have

TABLE 10.1   Outcomes/Conclusions of Selected, Important Studies on Reirradiation

Setting	Outcomes/Conclusions	Authors/Study
Retrospective, milestone study, n = 45. Unresectable locally or regionally recurrent disease. Hydroxyurea + 5FU alternating with RT.	Reirradiation is an aggressive treatment but can yield long-term survival at 5 years. 2-year survival was 35% and 8%, respectively, with RT doses more or less than 58 Gy. Fatal complications were observed and grades 3–4 late toxicity was common.	Haraf, DJ, Weichselbaum RR, Vokes EE (1996)
Randomised phase III trial: salvage surgery alone versus postoperative reirradiation with chemotherapy, n = 130. Resectable recurrent disease.	Significant improvement in locoregional control; overall survival showed no difference; more grades 3–4 late toxicities (39% vs 11% at 2 years, 5 treatment-related deaths).	Janot F, de Raucourt D, Benhamou E et al. (2008)
Retrospective study, n = 103. Comorbidity was assessed with Charlson index, 46 patients had salvage surgery before reirradiation.	Comorbidity and organ dysfunction are important prognostic factors. A nomogram to predict probability of death within 24 months after reirradiation was developed.	Tanvetyanon T, Padhya T, McCaffrey J et al. (2009)
Prospective non-randomised single-institution study on using daily image guidance during IMRT. (n = 21 consecutive patients).	IMRT with daily image guidance: image-guided radiation therapy (IGRT) results in effective disease control with relatively low morbidity; no treatment related fatalities were observed. Tumour volume of utmost importance for tumour control probability.	Chen AM, Farwell DG, Luu Q et al. (2011)
Retrospective single-institution study during 15 years. Institution with large experience.	Reirradiation with IMRT either in a definitive fashion or after surgery can produce local tumour control and survival in selected patients. Treatment related toxicity is important. Prognostic factors are emerging.	Takiar V, Garden AS, Da M et al. (2016)

been published using reirradiation as a salvage modality. It is important to understand that many of these earlier series were published where patients were treated with techniques currently considered obsolete; radiotherapy was delivered with standard, lateral-opposed, single wedge paired, or oblique fields targeting the gross tumour plus generous margins. Even if 3D-CRT based on CT was recommended, it was not required. Only three studies with a prospective randomised design were performed with these techniques, two for patients with unresectable disease and one for patients with reirradiation after attempted salvage surgery (see Table 10.1). The two studies for patients with unresectable, recurrent tumours were both closed prematurely because of poor accrual and results have not been reported.

10.3 OUTCOME WITH REIRRADIATION-LOCOREGIONAL CONTROL AND SURVIVAL

Almost two decades ago, the Chicago group presented data to imply that reirradiation to a previously irradiated area in the head and neck was feasible (Haraf, Weichselbaum & Vokes 1996). In this milestone study, the treatment schedule consisted of protracted radiotherapy, delivering 60 Gy over 11 weeks together with 5FU and hydroxyurea. Their experience suggested that reirradiation together with chemotherapy was feasible and could achieve long term tumour control in some patients even if late toxicity could be substantial. Similar results were reported in subsequent studies; in one of the reports studying reirradiation of unresectable disease 169 patients were treated and an overall survival at 2 and 5 years, respectively. was found to be 21% and 5%. In one of the two studies with a prospective, randomised phase III design, performed by the French GORTEC group, 57 out of the 160 planned patients were randomised between weekly single agent methotrexate or six cycles of reirradiation (5 × 2 Gy/fraction/cycle) and concurrent hydroxyurea/5FU with a 9-day inter-cycle rest (Tortochaux et al. 2011). No differences were found in 1- or 2-year overall survival, with 22% and 23%, respectively, alive at 1 year. The goal of the study was to evaluate the potential benefit of concurrent reirradiation with chemotherapy versus a single chemotherapeutic agent, which is one standard treatment with a palliative intention only. Slow recruitment of patients ended in a premature end to this study. However, there were no indications of an improvement in overall survival with the reirradiation schedule.

So far, there have not been published other randomised studies to indicate optimal approaches for irradiation of patients with recurrent or second primary tumours in previously irradiated areas in the head and neck. The Radiation Therapy Oncology Group (RTOG) launched a phase II study comparing concomitant chemotherapy and reirradiation with three cisplatin-based standard chemotherapy regimens (RTOG 0421). Also, this study was closed because of bad recruitment and results have not been reported. Therefore, evidence for prescribing reirradiation as a curative treatment for patients with unresectable disease without prior salvage surgery comes from retrospective and phase II trials (see Table 10.1).

However, reviewing the reports where modern radiotherapy techniques have been used (IMRT, SBRT, brachytherapy or protons), it can be concluded that at 2 years, overall survival rates between 10% and 30% can be expected and that 25% to 30% of patients will have a locoregional tumour control. Side effects were serious with grades 3–4 toxicities in up to 40% of patients and with almost 10% of patients dying from effects related to the treatment. The rather great spread in outcomes in all these often-small treatment series depends primarily on different selection criteria for including patients for reirradiation and also for the diversity of chosen reirradiation schedules, total radiation dose, fractionation, definition of target volume, addition of drugs concomitantly or as induction therapy, radiation technique, etc.

Therefore, conclusions concerning optimal reirradiation schedules certainly cannot be drawn: no randomised comparisons have been undertaken and the phase II studies conducted and reviewed show a great heterogeneity both with regard to patient eligibility criteria and therapeutic decision-making.

For patients reirradiated after attempted radical salvage surgery, one prospective phase III study has been published, comparing salvage surgery alone with postoperative reirradiation combined with chemotherapy (Janot et al. 2008). In this report, comprising of 130 randomised patients from French and Belgian centres, significant improvements were seen in locoregional tumour control. Nevertheless, overall survival did not differ because of more treatment-induced deaths, distant metastases, and second primaries among the reirradiated patients. Reirradiation was also followed by a significant increase of grades 3 and 4 late toxicities, 39% versus 10% at 2 years, and five patients died because of treatment-related causes in the reirradiated group.

Some smaller retrospective or prospective studies have been presented in the literature, in general supporting the results and conclusions drawn from the phase III study. In general, treatment results are superior with reirradiation after previous surgical salvage than for patients with unresectable disease. In one study, surgical resection before irradiation was independently associated with improved overall survival, progression free survival and local tumour control: a 3-year disease free survival rate of 51% vs 19% was reported for patients who had previous surgery compared to patients who did not (Salama et al. 2006). Other reports in the literature support the findings in this study. The reasons for this observation could be both clinical and have radiobiological foundations. Patients referred for surgery as a primary salvage procedure probably have lower tumour burdens from the beginning, even further reduced after surgery. Another reasonable hypothesis could relate to a better vascularisation in the target area after resection of scar tissue, making residual tumour cells less hypoxic and, therefore, more sensitive to ionising radiation.

In one recent report from an institution with profound experience to treat head and neck cancer patients, where the follow-up period of treatment outcome is rather long and where consistency concerning treatment schedule is reasonably good, important observations can be made regarding patient selection (Takiar et al. 2016). In their group of patients, 206 were reirradiated and of those with squamous cell histology, 80 patients had upfront surgery followed by reirradiation and 93 received definitive reirradiation only. Patients with recurrent tumours in the oral cavity, skin, parotid, or neck were more likely to receive salvage surgery before irradiation whereas those with recurrences in the nasopharynx, oropharynx, and retropharyngeal nodes were more likely to have definitive reirradiation. No significant differences were registered with respect to age, performance status, disease-free interval, interval between radiation therapy courses, initial radiation therapy dose, reir-radiation volume or chemotherapy use between surgical and nonsurgical patients. In this outcome report, median survival for patients with definitive reirradiation was 27.7 months and 22.8 months, respectively, for those who received adjuvant reirradiation after surgery, which is a nonsignificant difference. Locoregional failure was observed in 40% of patients and 22% failed distantly. Common distant failure sites were the lungs, mediastinum, both lungs and mediastinum and brain.

10.4 REIRRADIATED VOLUME

The volume of irradiated volume is dependent on the accuracy of defining extent of tumour, the size of the recurrent tumour, chosen margins for safety (GTV, CTV and PTV, respectively) and the irradiation technique. Undoubtedly, imaging plays an essential role in the workup of these patients and for the subsequent planning of the radiation treatment. CT is the mainstay of workup but MRI displays several advantages over CT, such as superior, soft-tissue contrast, and is less susceptible to artefacts from dental amalgam.

The role of PET/CT is still somewhat controversial for the treatment planning process even if its role in diagnosis and for monitoring of treatment response is emerging. However, in some centres, PET/CT is a routine procedure for the treatment planning of all patients who are candidates for radiotherapy, both for primary tumours and for recurrences. At least, when SBRT was chosen as the technique for the treatment of recurrent head and neck cancer, PET/CT planning had an impact on both failure-free survival and on combined overlap/marginal failure-free survival with significant improvements in those patients undergoing this procedure (Wang et al. 2012).

An accurate delineation of target volume is of the utmost importance for all patients having their primary tumour treated with curative radiotherapy. It is even more so for patients who are candidates for reirradiation of a recurrent tumour considering the high risk of serious toxicities. All measures, including physical examination, should, therefore, be undertaken to reach precision considering the extent of tumour mass. The margin from GTV to CTV, aiming at covering potential microscopic disease, is not defined in most reports but is described as 'narrow' or 'tight', typically 3–5 mm. Most reirradiation experiences have targeted only the recurrent gross disease without elective nodes in the volume. What could be considered appropriate in this respect can be disputed from different scenarios. In patients with an isolated local recurrence and with no tumour found in the neck at the time of recurrence and in whom at the time of the primary radiation, the neck was also N0 and then irradiated electively, the risk of occult tumour spread in this region is considered low, with no justification for retreatment of the neck (see Figure 10.1). On the other hand, in patients with an isolated local recurrence, who were initially treated for node-positive disease, elective treatment of the neck could be indicated (Solares et al. 2005).

Patients with small treatment volumes are reported to have better treatment outcomes. In one study, tumour volume was the only significant prognostic factor (De Crevoisier et al. 1998). More recently, in another study, it was found that patients with a tumour volume of less than 27 cm^3 had a 2-year local control rate of 80% and it was also concluded that in

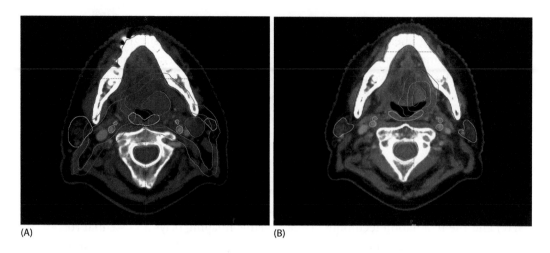

(A)   (B)

FIGURE 10.1   Target outlines for the primary base of tongue cancer **(A)** and the outlines for the recurrent tumour **(B)**.

patients with tumour volumes larger than 60 cm^3, reirradiation should be carefully considered (Chen et al. 2011).

The techniques used for reirradiation of patients are, as mentioned before, in the modern era, IMRT (complemented with IGRT), SBRT, IMPT and brachytherapy. Brachytherapy plays an important role for irradiation of these patients. In fact, the use of reirradiation in head and neck carcinoma patients was initially restricted to brachytherapy. Its usefulness has been widely reported with quite impressive outcomes, 5-year local control rates of 57%–69% and 5-year overall survival rates of 14%–40% (Langlois et al. 1988; Strnad, Geiger & Lotter 2003). Brachytherapy can be applied as monotherapy, in combination with external-beam radiotherapy or also combined with surgery as an intraoperative procedure for recurrent tumours in oral cavity, oropharynx, nasopharynx and lymph nodes and displays advantages over external beam radiotherapy because of its superior rapid dose fall-off in adjacent normal tissues. In reality, brachytherapy is restricted to highly experienced institutions. Evidence is now accumulating, however, that IMRT, like brachytherapy, could improve the outcome of reirradiation of head and neck cancer patients owing to its similar ability to offer highly conformal target coverage and limit dose to uninvolved normal tissues. IGRT can help to reduce repositioning errors and can be used to monitor the treatment region and adapt, if necessary, dose distribution to the possibly changing target and organs at risk during the radiation schedule.

In a report from Memorial Sloan Kettering Cancer Center, patients who underwent IMRT had dramatically superior locoregional control at 2 years compared with those treated by non-IMRT techniques, 52% vs 20%. Additionally, acute and late grade 3 or greater toxicity rates of 23% and 15%, respectively, compared favourably with those from historical controls (Lee et al. 2007). Similar results have been reported by other authors.

SBRT has been used in some centres, applied either in one single fraction as 'radio-surgery' or fractionated. One of the most important advantages with this technique is the short total treatment time, with the possibility to overcome the negative impact of accelerated repopulation of tumour clonogens during treatment time. Another important mechanism for increased cytotoxicity with high doses per fraction could also relate to an increased damage of the microvasculature in tumour tissue. Using the Cyber-Knife system, the authors reported an 80% response rate after a dose of 30 Gy (range 18–40 Gy) in 3–5 fractions. A 2-year survival of 30.9% and a treatment-related death rate of 2.9% was reported (Roh et al. 2009). In another study of 65 patients a similar dose and fractionation was used and applied with SBRT. A median reirradiation dose of 30 Gy in 2–5 fractions was delivered. Of the 38 patients treated with a curative intent 2-year, overall survival and locoregional control rates were 41% and 30% respectively, and 11% experienced severe toxicity with 1 patient's death attributed to the reirradiation (Unger et al. 2010).

A recent national survey has been undertaken in the United States to evaluate practice patterns related to salvage therapy of recurrent head and neck cancer. The questionnaire mailed to the physicians consisted of two clinical case scenarios followed by 82 clinical questions and four demographic questions. Among those clinicians recommending reirradiation with a curative intent for the two patients, conventionally fractionated IMRT was the most common treatment of choice followed by SBRT. SBRT was more likely to be recommended among physicians working in academic centres compared to other practitioners (Rwigema et al. 2017). It is clear that improvements are seen with the use of IMRT and SBRT with respect to locoregional control and toxicity patterns.

In a recent retrospective study including 414 patients from eight academic centres, IMRT and SBRT were compared for patients undergoing reirradiation for a tumour appearing within a volume previously irradiated to >40 Gy and then reirradiated with either IMRT to >40 Gy or SBRT delivered in 1–5 fractions of >5 Gy per fraction (Vargo et al. 2017). This study, which is to date the largest comparative study of these two modern, well-established reirradiation modalities, suggested that both approaches are relatively safe and compare favourably with historical studies of reirradiation. However, the authors conclude, also with a focus on future studies, that SBRT given in doses of >35 Gy has potential advantages with lower life-threatening acute tissue effects, potentially favouring the addition of novel systemic therapies especially for smaller tumours and also for poor prognosis patients, the latter patient group defined by having their reirradiation <2 years from the previous course and having significant organ dysfunctions (Vargo et al. 2017).

Proton therapy offers advantages for normal tissue protection owing to its sharp fall-off of dose characterised by the Bragg peak. Its role in clinical practice is emerging. In a recent study of proton reirradiation of 85 patients with oropharyngeal cancer, 39% given postoperatively and 61% as definitive radiotherapy, the 1-year locoregional control was 75% and overall survival 65%. Grade 3 or greater late skin and dysphagia toxicities were reported in six (8.7%) and four (7.1%) of patients, respectively. Two patients succumbed from treatment-related bleedings (Romesser et al. 2016).

10.5 REIRRADIATION DOSE AND FRACTIONATION, PROGNOSTIC FACTORS

In trying to analyse the literature in order to sort out prognostic factors for locoregional tumour control and survival, one is hampered by the limited sample size of each study, the inclusion of various treatment schedules, and often inadequate, short and inconsistent follow-up of patients. However, some information of importance can be drawn, notwithstanding the need for careful evaluation and further validation of data. In a retrospective study of 103 patients, a reirradiation dose below 50 Gy was an adverse factor for survival. Among other independent prognostic factors, they defined interval from previous irradiation, recurrent tumour stage, tumour bulk at reirradiation to be significant. They also found organ dysfunction and comorbidity, as evaluated and assessed by Charlson index to best highly prognostic. These findings enabled the authors to develop a nomogram to predict the probability of death within 24 months after reirradiation of patients with recurrent head and neck cancer (Tanvetyanon et al. 2009).

From the University of Chicago Pritzker School of Medicine, where the original milestone study came, supporting reirradiation of head and neck cancer patients as a potentially curative modality, a long-term follow-up showed radiation dose to be an independent factor for overall survival, progression-free survival and local control (Salama et al. 2006). Patients receiving a dose >58 Gy had a 3-year overall survival rate of 30% compared with 6% for those receiving a lower dose. While other authors report similar results, it must be pointed out that patients selected for higher radiation doses are mostly in a relatively good general health condition with a probably inherently better prognosis (Emami et al. 1987; Platteaux et al. 2011). Some authors find that the longer the time interval prior irradiation, the better the locoregional control and survival (Spencer 1999, 2001). It has also been suggested that the longer the time between the radiotherapy courses, the lower the risk of developing serious late effects. It has been suggested to be a higher risk of grade 3–4 toxicities in patients with an interval between the courses of <1 year (Chen et al. 2011).

No conclusions can be drawn with respect to fractionation of dose. Experimental studies as well as experiences from irradiation of radiotherapy-naïve patients would suggest hyperfractionated regimens to be the most effective to spare late-reacting tissues in the vicinity of the target. However, no apparent advantages have emerged with these regimens with regard to either tumour control or toxicity. In most reports, the acute toxicity profile seems similar to what is seen during the initial course of radiotherapy or even less intensive, probably because of a smaller target volume, as used in reirradiated patients (see Figure 10.1). When dose/fractionation schedules have been chosen for these patients, there is a balance to be considered between the inconvenience of protracted, hyperfractionated schedules versus the risk of increased late toxicity with short, hypofractionated courses.

Since the fraction size is of major concern for severe late toxicity, stereotactic body irradiation is an attractive option for head and neck radiotherapy because it is delivered with very high precision and a high dose per fraction in a small number of fractions leading to highly conformal dose delivery. In one study presenting long-term follow-up data of 184 patients, of whom 120 patients were reirradiated with stereotactic radiosurgery, there were serious late effects in 59 patients, the most common being temporal lobe necrosis in 15 patients (Owen, Iqbal & Pollock 2014). However, toxicity was reported to be very acceptable in a multi-institutional phase II study from France where SBRT was combined with weekly cetuximab. Caution was stressed not to reirradiate large neck nodes invading the vessels of the neck or spinal cord; it was recommended not to irradiate more than one-third of the carotid artery and avoid treating patients with tumours with skin infiltration (Lartigau et al. 2013).

In the American national survey of practice patterns for reirradiation of recurrent head and neck cancer, mentioned above, the most common dose was 60 Gy for these patients and the preferred fractionation scheme was conventional (72%) compared with hyperfractionated, twice-daily schemes (27%) (Rwigema et al. 2017).

10.6 REIRRADIATION WITH DRUGS

It could be hypothesised that the benefit of concurrent chemotherapy and also cetuximab in a reirradiation setting could be similar to what is seen in their respective large, randomised studies in the setting of primary radiotherapy. Whether the addition of systemic agents really improve effectiveness of the reirradiation is not known. In recent studies, both for patients in the postoperative and definitive setting, excellent locoregional control and overall survival was described using a well-defined radiotherapy protocol without any chemotherapy (Langendijk et al. 2006).

Different chemotherapy regimens have been used in the literature, but most have been platinum based or consisted of 5FU-hydroxyurea as was originally described by the pioneering studies from the Chicago group. No conclusions can be drawn, however, concerning their relative efficacy because of the retrospective nature of the reports and diversity of different study parameters. When cetuximab was substituted for a regimen consisting of conventional cytostatic drugs, the toxicity profile of treatment differed significantly. In one study, cetuximab was given in standard doses concomitant with SBRT, 5×8 Gy delivered every other day in 35 patients with recurrent tumours. The outcome of this treatment was matched with 35 other patients, retreated with SBRT alone. The matching procedure was performed with respect to age, sex, performance status prior therapy, previous radiation dose, interval to recurrence, as well as recurrent disease characteristic such as site, size and presence of systemic metastases. At 2 years, local control rates were 33.6% for the SBRT-alone group and 49.2% for the cetuximab-RT group and with 2-year survival rates of 21.1% and 53.3%, respectively. On multivariate analysis, cetuximab was one of the significant predictors for improved survival. There were no treatment-related deaths and the incidence of grade 3 late effects were 3% and 6%, respectively (Heron et al. 2011).

In the American national survey of practice patterns, 69% of oncologists favoured reirradiation with concurrent chemotherapy. The choice of agent(s) included single-agent cisplatin (23%), cetuximab (17%), carboplatin/paclitaxel (14%), and carboplatin (13%). A significantly higher proportion of academic physicians in this survey recommended reirradiation schedules without concurrent chemotherapy compared with their colleagues working in community-based or hybrid settings (44% vs 16%) (Rwigema et al. 2017).

10.7 NORMAL TISSUE TOXICITY

When the primary tumour in the head and neck is irradiated with a curative intention, there are several tissues and organs to consider in order to avoid effects that can negatively influence the patients' quality of life. Direct lethal and sublethal damage of tissues, along with immune system impairments, can contribute to such acute and late side effects of great importance. Such effects most often relate to mucositis, pain, loss of taste, impaired salivation (xerostomia), swallowing dysfunction (dysphagia), soft tissue necrosis, problems to open mouth (trismus), mandibular osteoradionecrosis, laryngeal oedema, skin reactions, temporal lobe necrosis, myelitis, and second malignant tumours. Acute toxicity, most often mucositis with oral and oropharyngeal pain and loss of taste, can be troublesome for the patient during the time of irradiation but can be adequately managed with analgesics and sometimes a supported nutrition; they will most often subsequently subside completely. Of greater concern are the late effects induced by the radiation. With modern radiotherapy techniques, the scenario has definitely changed during the last decades and optimisation procedures of dose distributions can often spare many tissues the influence of a damaging radiation dose, at least from grades 3 and 4 toxicity. However, problems with xerostomia and dysphagia still exist; they are sometimes categorised as grade 3 or 4 toxicity in some patients and also troublesome osteoradionecrosis, although in a low frequency, is also reported.

When reirradiation is decided for a patient, the fear is, of course, to pass the tolerance limits of all tissues ("tissue constraints"). If the radiation tolerance within a certain volume of an organ has already been exceeded during the primary therapy, intuitively, no further radiation dose should be given to this volume. However, when analysing data from reports in the literature, the observed incidence of normal tissue injury was clearly lower than expected in the reirradiation series, indicating at least partial long-term recovery of tissues in the head and neck area. Some authors, therefore, suggest that cumulative biologically equivalent doses of up to 130 Gy in 2 Gy fractions are safe (Nieder, Milas & Ang 2000; Kao et al. 2003).

The priority of sparing critical normal structures when planning reirradiation is now changing from what was considered most important when planning the primary treatment and focus is primarily on avoiding deadly threatening complications in the patient. The spinal cord was recently ranked as the most important structure to spare, followed by the brainstem, the carotid artery, the mandible, and the larynx. The most feared toxicity of all was the carotid artery blowout syndrome, followed by soft tissue necrosis, osteoradionecrosis, mucosal ulceration, and laryngeal oedema, respectively (Rwigema et al. 2017). In a systematic review of published data on carotid blowout, there was an incidence of 2.6% at a median time of 7.5 months post-reirradiation and 76% of these events were fatal. Both hyperfractionation protocols, hypofractionation protocols and protocols using conventional fractionation have been used but a heterogeneous patient population precludes from definitive conclusions about the impact of fractionation.

Some investigations find dysphagia to be the most frequently reported late radiation-induced morbidity in up to 50% or more among patients with a second course of radiation

(Langendijk et al. 2006; Choe et al. 2011). Mandibular osteoradionecrosis is observed at a rate ranging from 0 to 7% of reirradiated patients, a rate that has improved significantly with modern radiotherapy techniques (Sulman et al. 2009; Shikama et al. 2013).

Also, other chronic reactions can occur that are influenced by the localisation of the recurrent tumour mass. Thus, toxicity of varying grades has been described such as brain necrosis, blindness, pharyngocutaneous fistulas, palatal fibrosis, trismus, and chondronecrosis of the larynx.

GENERAL CONCLUSIONS

1. Reirradiation is the only treatment that is potentially curative for patients with recurrent unresectable squamous cell carcinoma of the head and neck.

2. Reirradiation can be dramatically toxic and devastating outcomes have been reported with treatment-related deaths.

3. Patient selection for reirradiation is probably of utmost importance with a careful assessment of life expectancy, evaluation of comorbidities, performance status, prior radiotherapy sequelae, e.g. carotid stenosis, soft-tissue fibrosis, osteoradionecrosis, and other severe toxicity.

4. The chance of tumour control within the reirradiated area is higher with a dose of at least 60 Gy.

5. The longer the interval to prior radiotherapy, the better the prognosis; an interval of at least 1 year could be advantageous.

6. Advanced radiotherapy techniques should be used, such as IMRT, SBRT, brachytherapy, or protons.

7. The reirradiated area should have secure margins around the diagnosed tumour.

8. No certain data exist to stipulate that the addition of chemotherapy improves tumour control or has an impact on severe toxicity.

Appendix: Radiobiological Modelling

A.1 HISTORICAL PERSPECTIVE

Various models have been proposed to describe the response of tissues to fractionated irradiation and the relationship to time, dose, and fractionation pattern. Some of these models, like the Strandqvist power law (Strandqvist 1944), the nominal standard dose, NSD (Ellis 1969), the cumulative radiation effect, CRE (Kirk, Gtay & Watson 1971), and the time-dose-and-fractionation tables of Orton and Ellis (1973) were derived from early clinical observations of normal tissue effects and described isoeffective schedules, but they were, in essence, empirical models with limited extrapolation power.

The development of in vitro cell survival experiments in the 1950s (Puck & Marcus 1956; Elkind & Sutton 1960), the successful transplantation of tumour cells between animal hosts and experiments showing that in vitro cell survival led to better understanding of the in vivo dose-effect relationship, (Alper 1973) shifted the focus in radiobiology to a more mechanistic approach that related radiobiological effects like tumour and normal tissue response to the survival or inactivation of cells. Consequently, radiobiology research became increasingly concerned with cell survival curves and their characterisation (Fowler 2006). Consequently, several models have been proposed to describe the survival curves for various cell lines and even tissues.

Among these models, a prominent place is taken by target theory models originally introduced by Lea (1946) to describe the mechanism of radiation mediated cell kill. The basic concept of target models was that cell killing depends on the energy deposition events above a certain threshold (called "hits") taking place in near proximity of critical structures in the cells (called "targets"). Various versions of the target models have been proposed, depending on the number of targets that had to be inactivated in the cells and the number of hits required for this purpose. Much hope was put into relating the number of hits to biologically relevant features; for example, a hit number of 2 might represent damage to the two strands of the DNA (Fowler 2006). Subsequent experiments showed very large ranges of values for the target models parameters, and the realisation that one critical lesion rather than many targets hit may be responsible for lethality, diminished initial excitement, and, eventually, led to their abandonment. However, these models are particularly interesting due to their lengthy employment in experimental practice.

The identification of the DNA as a target for radiation action and increasing knowledge on the repair of DNA damage led to the next radiobiological breakthrough by bringing together DNA damage and cell survival (Fowler 2006). Thus, it was thought that it is not a mere coincidence that the mathematically simple linear-quadratic expression used to describe the induction of chromosome lesions as a function of dose since the early 1940s (Lea & Catcheside 1942; Lea 1946) could also be used to describe the fractionation effects in tissues, as attempted by Fowler and Stern (1960, 1963). Further evidence brought about by the fraction-effective (Fe) plots of Douglas and Fowler (1976) eventually led to the adoption of the linear quadratic (LQ) model as the fundamental model in clinical and experimental radiobiology.

A.2 THE LQ MODEL

The LQ model has been proposed since the 1940s to explain the dose dependence on radiotherapy (Gray 1944). Nevertheless, much later, it has found great use for iso-effective comparisons of head and neck cancer treatments and for designing new schedules aimed at increasing tumour effect while minimising the effects in normal tissues.

Several theories have been proposed to describe the model in mechanistic terms. One of these theories, the molecular theory of cell survival (Chadwick & Leenhouts 1973), used a radiobiological approach to calculate the probabilities of lesion production and interaction at the level of the DNA molecule. Another, almost concomitant theory, the theory of dual radiation action (Kellerer & Rossi 1974), used microdosimetric approaches to calculate the lethal energy deposition events resulting from single particle tracks or from interaction of radiation tracks in targets in the cell. Interestingly enough, both theories led to the same expression of the effect of radiation (Equation A.1).

$$\text{Effect} = \alpha D + \beta D^2 \tag{A.1}$$

where Effect could be for example the production of lethal DNA lesions, D is the radiation dose and α and β are parameters related to the lethal energy depositions or lesions created from single or multiple events, respectively.

It is interesting to note that general models to describe the induction and repair of lesions like the lethal and potentially lethal model (Curtis 1986) could be reduced to the LQ formulation for the most common applications in radiotherapy and this has contributed to the longevity of the LQ model in clinical and experimental radiotherapy.

Cell survival fraction (SF) after dose D according to the LQ model could be obtained from Equation A.1 as given by Equation A.2.

$$SF = \exp(-\alpha D - \beta D^2) \tag{A.2}$$

From the perspective of the expression in Equation A.2, the effect in Equation A.1 is often described as the log cell kill after dose D (Fowler 2006).

The LQ model could also be used to describe the effects of fractionated delivery of radiation. Thus, the effect of n fractions of dose d, well separated to avoid the effects of incomplete repair, is given by Equation A.3.

$$\text{Effect} = n(\alpha d + \beta d^2) \qquad (A.3)$$

Rearranging the terms in Equation A.3 the fractionated effect as a function of the total dose D = nd could be obtained (Equation A.4).

$$\text{Effect} = D(\alpha + \beta d) \qquad (A.4)$$

The expressions in Equations A.1 and A.4 could be further developed to include the effects of repair of DNA lesions. Thus, according to the incomplete repair model (Thames 1985), the effect of continuous irradiation delivering radiation with doserate $\dot{d}$ in time t is given by Equation A.5.

$$\text{Effect} = \alpha \dot{d} t + g(t)\beta (\dot{d} t)^2 \qquad (A.5)$$

where $g(t)$ is a function describing the repair taking place in time t. Assuming that repair is an exponential process with repair constant μ, the g term is given by Equation A.6.

$$g(\mu t) = 2\frac{\left[\mu t - 1 + \exp(-\mu t) \right]}{(\mu t)^2} \qquad (A.6)$$

Equations A.5 and A.6 have found great use in calculating the effectiveness of brachytherapy schedules. However, brachytherapy is little used in the management of head and neck cancers compared with external beam radiotherapy.

In the case of fractionated delivery of radiation when intrafraction repair could be neglected and the focus is on interfraction repair, the effect is given by Equation A.7.

$$E = n\left\{ \alpha d + \left[1 + h(t) \right]\beta d^2 \right\} \qquad (A.7)$$

where $h(t)$ is a repair function depending on the time interval between two successive fractions, as described by Equation A.8.

$$h(t) = \frac{2}{n}\frac{\exp(-\mu t)}{1 - \exp(-\mu t)}\left[n - \frac{1 - \exp(-n\mu t)}{1 - \exp(-\mu t)} \right] \qquad (A.8)$$

In the context of accounting for the effect of incomplete repair, Nilsson, Thames, and Joiner (1990) later proposed a generalised formulation to describe the effect of fractionated low-dose rate irradiation, although this is a situation seldom seen in modern clinical practice. Nevertheless, Equations A.1–A.8 give the classic expressions of the LQ model to describe the effect of ionising radiation in living systems.

The LQ model could be used to describe the response of tumours and normal tissues by calculating the corresponding tumour control probabilities (TCP) or normal tissue complication probabilities (NTCP). However, the adoption of these concepts and the underlying LQ model for clinical radiotherapy was hampered by the lack of reliable alpha and beta parameters to describe the various responses of the normal tissues.

An important step toward the adoption of the LQ model for clinical fractionation was made by Barendsen (1982), who proposed expressing the effect of a fractionation schedule in terms of the dose given in infinitely small doses per fraction that leads to the same effect for the same tissue. This dose, named extrapolated tolerance dose (ETD) in the original publication (Barendsen 1982) to quantify the influence of dose fractionation on the tolerance of normal tissues, could be obtained by dividing the expression in Equation A.4 by the α parameter of the LQ model (Equation A.9).

$$\text{ETD} = \text{Effect}/\alpha = D(1 + \beta d/\alpha) \qquad (A.9)$$

Rearranging the terms in Equation A.9 leads to the expression in Equation A.10.

$$\text{ETD} = D[1 + d/(\alpha/\beta)] \qquad (A.10)$$

It is important to note that Equation A.10 only includes the total dose D, the dose per fraction d and the ratio α/β of the parameters of the LQ model. Barendsen (1982) denoted the term in square brackets in Equation A.10 relative effectiveness (RE). The RE (Equation A.11) depends only on the dose per fraction d and the α/β ratio and could, therefore, be regarded as a quantity specific to a certain fractionation schedule through the d term and the tissue type through the α/β ratio.

$$\text{RE} = 1 + d/(\alpha/\beta) \qquad (A.11)$$

Consequently, the ETD of any fractionation schedule could be obtained as the product between the total dose D and RE of the fractionated schedule on a particular tissue type (Equation A.12).

$$\text{ETD} = D \cdot \text{RE} \qquad (A.12)$$

This was an important development as α/β ratios were derived with reliable confidence intervals for many normal tissues and even tumours from split dose experiments (Thames et al. 1982; Williams, Denekamp & Fowler 1985; Thames & Hendry 1987). Consequently, the calculation of isoeffectiveness of fractionated schedules used clinically was very much facilitated.

It was soon realised that the ETD concept could be applied not only to normal tissues but also to any biological damage, including cell kill; this led to it being renamed extrapolated response dose (ERD), a term that is still in use in some countries (Fowler 2006). The ERD provided a means to compare various fractionated schedules varying both the total dose and the dose per fraction.

According to this approach, two fractionation schedules that lead to the same effect have the same ERD. Consequently, one could write (Equation A.13):

$$D_1 \cdot RE_1 = D_2 \cdot RE_2 \qquad (A.13)$$

Where the term to the left described the effectiveness of the fractionation schedule delivering total dose D_1 with dose per fraction d_1 and relative effectiveness $RE_1 = 1 + d_1/(\alpha/\beta)$ and D_2 is the total dose of fractionation delivered with dose per fraction d_2 and relative effectiveness $RE_2 = 1 + d_2/(\alpha/\beta)$. From Equation A.13, one could calculate the total dose with a certain fractionation with relative effectiveness RE_2, which will lead to the same effect as a schedule with relative effectiveness RE_1 (Equation A.14).

$$D_2 = D_1 RE_1 / RE_2 \qquad (A.14)$$

The ERD concept could also be used to calculate the effectiveness of treatments delivered with different fractionations. For example, let us assume that the treatment delivers dose $D_{elective}$ in $d_{elective}$ dose per fraction to the elective target volume, while the boost delivers D_{boost} in d_{boost} dose per fraction to the boost target volume, included in the elective target volume. Thus, the total ERD for the boost volume will be (Equation A.15).

$$ERD_{tot} = D_{elective} \cdot RE_{elective} + D_{boost} \cdot RE_{boost} \qquad (A.15)$$

These properties allowed increased flexibility over other approaches to account for the effects of fractionation, but ERD lacked a means to include the overall treatment time in the calculation and, therefore, to account for the effects of treatment gaps that were shown to lead to decreased response.

A.3 EFFECTS OF OVERALL TREATMENT TIME AND THE BIOLOGICALLY EFFECTIVE DOSE (BED)

The breakthrough for the clinical adoption of the LQ model was caused by the understanding of the dynamics of in-treatment repopulation illustrated by the so-called "dog-leg" curve (Withers, Taylor & Maciejewski 1988) and the incorporation of its effects in the expression of the LQ model. Thus, Travis and Tucker (1987) gave an expression for the effect of a total dose D given with a dose per fraction d in overall time T (Equation A.16).

$$Effect = D(\alpha + \beta d) - f(T) \qquad (A.16)$$

where f(T) is a term that describes the effect of repopulation occurring in time T. Assuming exponential repopulation stating at time T_k with a doubling time T_p, f(T) has the expression in Equation A.17.

$$f(T) = \ln(2) \cdot (T - T_k)/T_p \qquad (A.17)$$

Consequently, Fowler (1989) proposed new expressions for the ERD to account for overall treatment time (Equation A.18), resulting in a quantity that was subsequently called biologically effective dose or BED.

$$BED = D\left[1 + d/(\alpha/\beta)\right] - \ln(2)/\alpha \cdot (T - T_k)/T_p \qquad (A.18)$$

It is interesting to note that the derivation of the expression in Equation A.18 has been brilliantly described by Prof. Jack Fowler as the "seven steps to linear quadratic heaven" (Fowler 2006; Fowler, Dasu & Toma-Dasu 2015).

It is also important to recognise that the BED is, in essence, the same as the ERD with all its properties outlined in the previous section, but the expression in Equation A.18 provides a relationship between time, dose, and fractionation patterns that allows the calculation of isoeffectiveness of various fractionation schedules, including the effect of treatment gaps on tumour and normal tissue response. Furthermore, the BED values could be converted into biologically equivalent doses for a conventional fractionation (typically 2 Gy per fraction) by dividing them to the RE of the latter (Equation A.19). These doses are usually described as normalised total doses (NTD) or equivalent dose in 2 Gy fractions (EQD$_2$).

$$NTD = BED/RE_{2Gy} = BED/\left[1 + 2/(\alpha/\beta)\right] \qquad (A.19)$$

Acknowledging the absence of proliferation and repopulation for late reacting tissues and the differences in parameters between tissues, different expressions for the BED could be derived for various tissues of interest in head and neck radiotherapy: tumour, late-reacting normal tissues, and acute mucosal reactions. In this context, it has to be mentioned that Fowler (1989) proposed the use of subscripts for the BED to distinguish it from being a true physical dose and to identify the α/β used for the calculation. As $\alpha/\beta = 10$ Gy was considered representative for early reacting tissues and $\alpha/\beta = 2$ or 3 Gy for late reacting tissues, the BED would be noted BED$_{10}$ for most tumours and acutely reacting tissues and BED$_2$ or BED$_3$ for late-reacting tissues.

Taking these into account, the BED for tumours is given by the expression in Equation A.20.

$$BED_{tumour} = D\left[1 + d/(\alpha/\beta)\right] - \ln(2)/\alpha \cdot (T - T_k)/T_p \qquad (A.20)$$

Typical values for parameters for head and neck tumours are $\alpha/\beta = 10$ Gy, $\alpha = 0.3 - 0.35$ Gy^{-1}, $T_k = 21 - 32$ days, and $T_p = 2.5 - 3$ days. It is important to note that, in many cases, T_p is different from T_{pot}, the potential doubling time describing the birthrate of cells in undisturbed tumours in the absence of cell loss, and reliable calculations have been performed with the mentioned generic values. Furthermore, the overall treatment time T in Equation A.20 is calculated by assuming that the first fraction of radiation is delivered in day 0 of the schedule.

For late-reacting normal tissues, the BED is given by the expression in Equation A.21.

$$BED_{\text{late reactions}} = D\left[1 + d/(\alpha/\beta)\right] \tag{A.21}$$

Typical values for tissues of interest in head and neck radiotherapy are $\alpha/\beta = 2$ Gy for the central nervous system and $\alpha/\beta = 3$ Gy for most others tissues (Joiner & van der Kogel 2009). Acute mucosal reactions have a similar BED expression as tumours, but with various parameters (Equation A.22).

$$BED_{\text{acute reactions}} = D\left[1 + d/(\alpha/\beta)\right] - \ln(2)/\alpha \cdot (T - T_k)/T_p \tag{A.22}$$

Typical values for acute mucosal reactions to be used with Equation A.22 are $\alpha/\beta = 10$ Gy, $\alpha = 0.3 - 0.35$ Gy^{-1}, $T_k = 7$ days and $T_p = 2.5$ days.

Figure A.1 illustrates the variation of calculated BEDs for tumour and normal tissues reactions for conventionally fractionated schedules with 2 Gy per fraction.

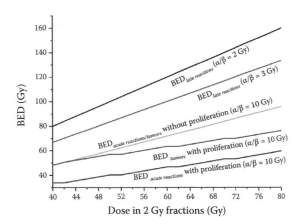

FIGURE A.1 Illustration of the BED values for tumour and normal tissue late and acute reactions for conventionally fractionated schedules with 2 Gy per fraction. The BED values for the late reacting normal tissues and tumours were calculated accounting or not for the effect of repopulation during the treatment time. The parameters used for the calculations of BED $_{\text{tumour}}$ are: $\alpha/\beta = 10$ Gy, $\alpha = 0.35$ Gy^{-1}, $T_k = 28$ days and $T_p = 2.5$ days. The parameters used for the calculations of BED$_{\text{acute reactions}}$ are: $\alpha/\beta = 10$ Gy, $\alpha = 0.35$ Gy^{-1}, $T_k = 7$ days and $T_p = 2.5$ days. The parameters used for the calculations of BED$_{\text{late reactions}}$ are: $\alpha/\beta = 2$ Gy and $\alpha/\beta = 3$ Gy, respectively.

A.4 EXAMPLES OF BED CALCULATIONS

Equations A.20–A.22 have been used for a long time for isoeffectiveness calculations in head and neck radiotherapy and have led to the introduction of many successful schedules. Thus, BED calculations could be used to search for treatment schedules that increase the BED_{tumour}, while keeping the corresponding BED values for normal tissues below their maximum tolerance (Fowler 2008; Fowler 2010; Fowler, Dasu & Toma-Dasu 2015).

For example, the schedule, which is usually associated with the maximum tolerance of late reacting normal tissues, 70 Gy in 2 Gy fractions over 7 weeks with 5 fractions per week, has the following BEDs.

$$BED_{tumour} = 70 \cdot [1 + 2/10] - \ln(2)/0.35 \cdot (46 - 28)/3 = 72.1 \, Gy \tag{A.23}$$

$$BED_{late \ reactions} = 70 \cdot [1 + 2/3] = 116.7 \, Gy \tag{A.24}$$

$$BED_{acute \ reactions} = 70 \cdot [1 + 2/10] - \ln(2)/0.35 \cdot (46 - 7)/2.5 = 53.1 \, Gy \tag{A.25}$$

The BED in Equation A.24 is considered the upper limit for the late effects. It should be mentioned that a similar upper limit for the acute mucosal reactions has been defined around 61 Gy (Fowler et al. 2003), which has not been exceeded in this case.

In Equation A.23, it has been assumed that the treatment started on a Monday and; therefore, the overall treatment time until the Friday when the last treatment fraction was given has been 46 days. This BED could be converted into the corresponding NTD by dividing it to the RE of the conventional fractionation with 2 Gy per fraction, RE = 1 + 2/10. Thus, NTD_{tumour} = 72.1/(1 + 2/10) = 60.1 Gy in 2 Gy fractions if there is no repopulation during the treatment.

If the treatment started on a Tuesday instead of a Monday, or if the patient loses 1 day of treatment due to the breakdown of the treatment machine, the treatment duration for the same schedule of 70 Gy in 2 Gy fractions would be 49 days. Under these conditions, the BED_{tumour} would become:

$$BED_{tumour} = 70 \cdot [1 + 2/10] - \ln(2)/0.35 \cdot (49 - 28)/3 = 70.1 \, Gy \tag{A.26}$$

while the BED_{late} will not change as it is independent of the overall treatment time.

Consequently, the corresponding NTD_{tumour} = 70.1/(1 + 2/10) = 58.4 Gy (in 2 Gy fractions). These calculations indicate that the addition of a weekend due to the delay in starting the treatment would be equivalent to a loss of 1.7 Gy, which is slightly less than a full fraction.

Let us look now at one of the most used fractionation schedules for head and neck tumours, delivering 68 Gy in 34 fractions of 2 Gy. If the treatment starts on a Monday or a Tuesday, the treatment duration is 45 days. For this schedule, the three BEDs of interest are:

$$BED_{tumour} = 68 \cdot [1 + 2/10] - \ln(2)/0.35 \cdot (45 - 28)/3 = 70.4\,Gy \qquad (A.27)$$

$$BED_{late\ reactions} = 68 \cdot [1 + 2/3] = 113.3\,Gy \qquad (A.28)$$

$$BED_{acute\ reactions} = 68 \cdot [1 + 2/10] - \ln(2)/0.35 \cdot (45 - 7)/2.5 = 51.5\,Gy \qquad (A.29)$$

All normal tissue BEDs are below the tolerance values and the NTD_{tumour} for this treatment is 58.6 Gy. This is slightly lower than the corresponding value for the schedule delivering 70 Gy in 2 Gy fractions over 7 weeks (see Equation A.23), but higher than the value from the same schedule delivered over an extra weekend (see Equation A.26). This illustrates the importance of overall treatment time in head and neck radiotherapy.

Let us assume that one introduces a gap of one week in the treatment. The treatment duration becomes 52 days instead of the original 45 days and the BED_{tumour} and $BED_{acute\ reaction}$ become:

$$BED_{tumour} = 68 \cdot [1 + 2/10] - \ln(2)/0.35 \cdot (52 - 28)/3 = 65.8\,Gy \qquad (A.30)$$

$$BED_{acute\ reactions} = 68 \cdot [1 + 2/10] - \ln(2)/0.35 \cdot (52 - 7)/2.5 = 46.0\,Gy \qquad (A.31)$$

Which correspond to an NTD_{tumour} = 54.8 Gy and an $NTD_{acute\ reactions}$ = 38.3 Gy. These values are 3.9 and 4.6 Gy lower than the corresponding NTD values for the original schedule. While the 4.6 Gy decrease for acute reactions might be regarded as beneficial for the quality of life of the patient during the treatment, the 3.9 Gy decrease in equivalent tumour response indicates that two additional fractions would be required for the treatment with a 1 week gap to achieve the effectiveness of the original schedule. It is, however, doubtful whether they could be delivered to maintain tumour response as the two extra fractions will increase the $BED_{late\ reactions}$ to 120 Gy, which is above the maximum tolerance of 116.7 Gy and would probably lead to a dramatic increase of the risk for normal tissue necrosis.

Let us assume instead that the 68 Gy in 2 Gy fractions treatment is delivered with six fractions per week in 38 days. This is the short DAHANCA treatment (Overgaard et al. 2003) and for this schedule the BED_{tumour} and $BED_{acute\ reaction}$ become:

$$BED_{tumour} = 68 \cdot [1 + 2/10] - \ln(2)/0.35 \cdot (38 - 28)/3 = 75.0\,Gy \qquad (A.32)$$

$$BED_{acute\ reactions} = 68 \cdot [1 + 2/10] - \ln(2)/0.35 \cdot (38 - 7)/2.5 = 57.0\,Gy \qquad (A.33)$$

Which correspond to $\mathrm{NTD}_{\text{tumour}} = 62.5$ Gy and $\mathrm{NTD}_{\text{acute reactions}} = 47.5$ Gy. These results indicate that the removal of one whole week from the treatment increases the corresponding NTDs for tumours and acute reactions by 3.9 and 4.6 Gy, respectively. The additional effectiveness of the schedule leads to improved tumour response at the expense of increased acute mucosal reactions as has indeed been seen when the schedule was compared clinically with the one delivered over 45 days. Nevertheless, it is important to note that the inclusion of one extra fraction each treatment week increased the effectiveness of the treatment, without impacting the effects in the late reacting normal tissues.

It is also interesting to analyse the effects of hyperfractionated schedules that exploit the differential sensitivity to fractionation of acute and late reacting tissues. One of the earliest such schedules was CHART, which delivered 36 fractions of 1.5 days, three times a day, over 11.5 days (Saunders et al. 2010). For this schedule, the three BEDs of interest are:

$$\mathrm{BED}_{\text{tumour}} = 54 \cdot [1 + 1.5/10] = 62.1 \, \mathrm{Gy} \tag{A.34}$$

$$\mathrm{BED}_{\text{late reactions}} = 54 \cdot [1 + 1.5/3] = 81.0 \, \mathrm{Gy} \tag{A.35}$$

$$\mathrm{BED}_{\text{acute reactions}} = 54 \cdot [1 + 1.5/10] - \ln(2)/0.35 \cdot (11.5 - 7)/2.5 = 58.5 \, \mathrm{Gy} \tag{A.36}$$

As the treatment duration in this case was shorter than the kick-off threshold for accelerated repopulation in tumours, the corresponding term disappeared from the BED expression for tumours. However, the term is still present for the BED describing the acute reactions as they have a much shorter kick-off threshold. These results indicate that while the schedule is less effective for tumours than the standard 68 Gy in 2 Gy fractions ($\mathrm{NTD}_{\text{tumour}} = 51.8$ Gy), the sparing of the late reactions is dramatic ($\mathrm{NTD}_{\text{late reactions}} = 48.6$ Gy), while the acute reactions are slightly more severe than those of the short DAHANCA schedule mentioned above ($\mathrm{NTD}_{\text{acute reactions}} = 48.8$ Gy).

Alternative schedules exist aiming to deliver a much higher effect to the tumour. One of these is the Radiation Therapy Oncology Group (RTOG) schedule delivering 68 fractions of 1.2 Gy twice a day over 45 days (Fu et al. 2000), for which the following BEDs could be calculated.

$$\mathrm{BED}_{\text{tumour}} = 81.6 \cdot [1 + 1.2/10] - \ln(2)/0.35 \cdot (45 - 28)/3 = 80.2 \, \mathrm{Gy} \tag{A.37}$$

$$\mathrm{BED}_{\text{late reactions}} = 81.6 \cdot [1 + 1.2/3] = 114.2 \, \mathrm{Gy} \tag{A.38}$$

$$\mathrm{BED}_{\text{acute reactions}} = 81.6 \cdot [1 + 1.2/10] - \ln(2)/0.35 \cdot (45 - 7)/2.5 = 61.3 \, \mathrm{Gy} \tag{A.39}$$

These values illustrate the potential of hyperfractionation. Thus, the schedule is isoeffective from the point of view of the late reactions with 68 Gy in 34 fractions, while the BED_{tumour} is almost 14% higher than the corresponding value from the same schedule. However, the $BED_{acute\ reactions}$ is very high, at the level of the tolerance value mentioned in the beginning of this section.

A somewhat milder alternative is the hyperfractionated EORTC schedule, delivering 70 fractions of 1.15 Gy twice a day over 46 days (Horiot et al. 1992). This schedule is characterised by the following BED values.

$$BED_{tumour} = 80.5 \cdot [1+1.15/10] - \ln(2)/0.35 \cdot (46-28)/3 = 77.9\,Gy \quad (A.40)$$

$$BED_{late\ reactions} = 80.5 \cdot [1+1.15/3] = 111.4\,Gy \quad (A.41)$$

$$BED_{acute\ reactions} = 80.5 \cdot [1+1.15/10] - \ln(2)/0.35 \cdot (46-7)/2.5 = 58.9\,Gy \quad (A.42)$$

Therefore, this schedule results in almost the same tumour effect, with comparable late effects and somewhat reduced acute effects.

It is interesting to note that TCP modelling of these schedules lead to predictions that were comparable with the outcomes observed in clinical studies (Hendry et al. 1996; Dale et al. 2002; Fowler, Dasu & Toma-Dasu 2015), but describing this type of modelling is beyond the purpose of this chapter. Further information on modelling approaches can be found in Dale and Jones (2007).

BED calculations could also be used to compensate for missed treatment days in radiotherapy (Hendry et al. 1996; Dale et al. 2002; RCR publication 2008) or at changes in fractionation during the treatment or when devising simultaneous integrated boost approaches (as illustrated by equations A.14 and A.15). As for the other fractionation schedules, it is important that all three components are calculated to ensure not only good tumour response, but also that the tolerances of normal tissues, acutely and late reacting, are not exceeded. Furthermore, the BED equations could be adapted to account for repair (Equations A.5–A.8), dose heterogeneity and volume effects (Hoffmann 2013), as well as for the effectiveness of radiation types other than photons (Dale & Jones 1999; Jones & Dale 2000; Jones, Carabe-Fernandez & Dale 2006; Dasu & Toma-Dasu 2008).

References

Ackerknecht, E. H. (1958). Historical notes on cancer. *Med Hist*, 2, 114–119.

Adams, J. and Kauffman, M. (2004). Development of the proteasome inhibitor VELCADE (bortezomib). *Cancer Invest*, 22, 304–311.

Adams, T. E., Epa, V. C., Garrett, T. P. et al. (2000). Structure and function of the type 1 insulin-like growth factor receptor. *Cell Mol Life Sci*, 57, 1050–1093.

Addeo, R., Caraglia, M., and Iuliano, G. (2016). Pembrolizumab: the value of PDL1 biomarker in head and neck cancer. *Expert Opin Biol Ther*, 16(9), 1075–1079.

Adelstein, D. J., Saxton, J. P., Lavertu, P. et al. (1997). A phase III randomized trial comparing concurrent chemotherapy and radiotherapy with radiotherapy alone in resectable stage III and IV squamous cell head and neck cancer: preliminary results. *Head Neck*, 19, 567–575.

Agrawal, N., Frederick, M. J., Pickering, C. R. et al. (2011). Exome sequencing of head and neck squamous cell carcinoma reveals inactivating mutations in NOTCH1. *Science*, 333, 1154–1157.

Agrawal, S., Awasthi, R., Singh, A. et al. (2012). An exploratory study into the role of dynamic contrast-enhanced (DCE) MRI metrics as predictors of response in head and neck cancers. *Clin Radiol*, 67(9), e1–e5.

Aguiar, P. N., Tadokoro, H., da Silva, G. F. et al. (2016). Definitive chemoradiotherapy for squamous head and neck cancer: cisplatin versus carboplatin? A meta-analysis. *Future Oncol*, 12(23), 2755–2764.

Ahlbom, H. E. (1935). Mucous- and salivary-gland tumours. A clinical study with special reference to radiotherapy, based on 254 cases treated at Radiumhemmet, Stockholm. *Acta Radiologica*, (Suppl 23).

Ahlbom, H. E. (1941). The results of radiotherapy of hypopharyngeal cancer at the Radiumhemmet, Stockholm, 1930 to 1939. *Acta Radiologica*, 22, 155–171.

Al Hajj, M., Wicha, M. S., Benito-Hernandez, A. et al. (2003). Prospective identification of tumorigenic breast cancer cells. *Proc Natl Acad Sci USA*, 100(7), 3983–3988.

Alber, M., Paulsen, F., Eschmann, S. M. et al. (2003). On biologically conformal boost dose optimization. *Phys Med Biol*, 48, N31–N35.

Alper, T. A. (1973). The relevance of experimental radiobiology to radiotherapy. Present limitations and future possibilities. *Br Med Bull*, 29(1), 3–6.

Alper, T. and Howard-Flanders, P. (1956). Role of oxygen in modifying the radiosensitivity of E. coli B. *Nature*, 178(4540), 978–979.

Al-Sarraf, M., LeBlanc, M., Giri, P. G. et al. (1998). Chemoradiotherapy versus radiotherapy in patients with advanced nasopharyngeal cancer: phase III randomized Intergroup study 0099. *J Clin Oncol*, 16, 1310–1317.

Ang, K. K and Sturgis, E. M. (2012). Human papillomavirus as a marker of the natural history and response to therapy of head and neck squamous cell carcinoma. *Semin Radiat Oncol*, 22(2), 128–142.

Ang, K. K., Chen, A., Curran, W. J. et al. (2012). Head and neck carcinoma in the United States: first comprehensive report of the Longitudinal Oncology Registry of Head and Neck Carcinoma (LORHAN). *Cancer*, 118(23), 5783–5792.

Ang, K. K., Harris, J., Wheeler, R. et al. (2010). Human papillomavirus and survival of patients with oropharyngeal cancer. *N Engl J Med*, 363(1), 24–35.

Ang, K. K., Zhang, Q., Rosenthal, D. I. et al. (2014). Randomized phase III trial of concurrent accelerated radiation plus cisplatin with or without cetuximab for stage III to IV head and neck carcinoma: RTOG 0522. *J Clin Oncol*, 32(27), 2940–2950.

Antonovic, L., Dasu, A., Furusawa, Y. et al. (2015). Relative clinical effectiveness of carbon ion radiotherapy—theoretical modelling for H&N tumours. *J Radiat Res*, 56(4), 639–645.

Antonovic, L., Lindblom, E., Dasu, A. et al. (2014). Clinical oxygen enhancement ratio of tumors in carbon ion radiotherapy: the influence of local oxygenation changes. *J Radiat Res*, 55, 902–911.

Apisarnthanarax, S. and Chao, K. S. (2005). Current imaging paradigms in radiation oncology. *Radiat Res*, 163, 1–25.

Arenz, A., Ziemann, F., Mayer, C. et al. (2014). Increased radiosensitivity of HPV-positive head and neck cancer cell lines due to cell cycle dysregulation and induction of apoptosis. *Strahlenther Onkol*, 190, 839–846.

Argiris, A., Bauman, J. E., Ohr, J. et al. (2016). Phase II randomized trial of radiation therapy, cetuximab, and pemetrexed with or without bevacizumab in patients with locally advanced head and neck cancer. *Ann Oncol*, 27(8), 1594–1600.

Argiris, A., Ghebremichael, M., Gilbert, J. et al. (2013). Phase III randomized, placebo-controlled trial of docetaxel with or without gefitinib in recurrent or metastatic head and neck cancer: an Eastern Cooperative Oncology Group Trial. *J Clin Oncol*, 31(11), 1405–1414.

Argiris, A., Karamouzis, M. V., Raben, D. et al. (2008). Head and neck cancer. *The Lancet*, 371, 1695–1709.

Argiris, A., Li, S., Savvides, P. et al. (2017). Phase III randomized trial of chemotherapy with or without bevacizumab (B) in patients (pts) with recurrent or metastatic squamous cell carcinoma of the head and neck (R/M SCCHN): survival analysis of E1305, an ECOG-ACRIN Cancer Research Group trial. *J Clin Oncology*, 35, 15.

Ask, A., Björk-Eriksson, T., Zackrisson, B. et al. (2005). The potential of proton beam radiation therapy in head and neck cancer. *Acta Oncol*, 44(8), 876–880.

Attner, P., Du, J., Näsman, A. et al. (2010). The role of human papillomavirus in the increased incidence of base of tongue cancer. *Int J Cancer*, 126, 2879–2884.

Bachar, G. Y., Goh, C., Goldstein, D. P. et al. (2010). Long-term outcome after surgical salvage for recurrent tonsil carcinoma following radical radiotherapy. *Europ Arch Otorhinolaryngol*, 267, 295–301.

Bachaud, J. M., Cohen-Jonathan, E., Alzieu, C. et al. (1996). Combined postoperative radiotherapy and weekly cisplatin infusion for locally advanced head and neck carcinoma: final report of a randomized trial. *Int J Radiat Oncol Biol Phys*, 36, 999–1004.

Bailet, J. W., Abemayor, E., Jabour, B. A. et al. (1992). Positron emission tomography: a new, precise imaging modality for detection of primary head and neck tumors and assessment of cervical adenopathy. *Laryngoscope*, 102, 281–288.

Bairati, I., Meyer, F., Gélinas, M. et al. (2005). Randomized trial of antioxidant vitamins to prevent acute adverse effects of radiation therapy in head and neck cancer patients. *J Clin Oncol*, 23(24), 5805–5813.

Bakst, R. L., Lee, N., Pfister, D. G. et al. (2011). Hypofractionated dose-painting intensity modulated radiation therapy with chemotherapy for nasopharyngeal carcinoma: a prospective trial. *Int J Radiat Oncol Biol Phys*, 80(1), 148–153.

Balermpas, P., Michael, Y., Wagenblast, J. et al. (2014). Tumour-infiltrating lymphocytes predict response to definitive chemoradiotherapy in head and neck cancer. *Br J Cancer*, 110, 501–509.

Balermpas, P., Rödel, F., Rödel, C. et al. (2016). CD8+ tumour-infiltrating lymphocytes in relation to HPV status and clinical outcome in patients with head and neck cancer after postoperative chemoradiotherapy: a multicentre study of the German cancer consortium radiation oncology group (DKTK-ROG). *Int J Cancer*, 138(1), 171–181.

Barendsen, G. W. (1982). Dose fractionation, dose rate, and isoeffect relationships for normal tissue responses. *Int J Radiat Oncol Biol Phys*, 8, 1981–1997.

Barnes, C. J., Ohshiro, K., Rayala, S. K. et al. (2007). Insulin-like growth factor receptor as a therapeutic target in head and neck cancer. *Clin Cancer Res*, 13, 4291–4299.

Baruah, P., Lee, M., Wilson, P. O. et al. (2015). Impact of p16 status on pro- and anti-angiogenesis factors in head and neck cancers. *Br J Cancer*, 113(4), 653–659.

Baumann, M. and Krause, M. (2010). CD44: a cancer stem cell-related biomarker with predictive potential for radiotherapy. *Clin Cancer Res*, 16(21), 5091–5093.

Baumann, M., Krause, M., and Hill, R. (2008). Exploring the role of cancer stem cells in radioresistance. *Nat Rev Cancer*, 8(7), 545–554.

Beasley, N. J., Wykoff, C. C., Watson, P. H. et al. (2001). Carbonic anhydrase IX, an endogenous hypoxia marker, expression in head and neck squamous cell carcinoma and its relationship to hypoxia, necrosis, and microvessel density. *Cancer Res*, 61, 5262–5267.

Becker, A., Kuhnt, T., Liedtke, H. et al. (2002). Oxygenation measurements in head and neck cancers during hyperbaric oxygenation. *Strahlenther Onkol*, 178(2), 105–108.

Begg, A. C., Haustermans, K., Hart, A. A. et al. (1999). The value of pretreatment cell kinetic parameters as predictors for radiotherapy outcome in head and neck cancer: a multicenter analysis. *Radiother Oncol*, 50, 13–23.

Beitler, J. J., Zhang, Q., Fu, K. K. et al. (2014). Final results of local-regional control and late toxicity of RTOG 9003: a randomized trial of altered fractionation radiation for locally advanced head and neck cancer. *Int J Radiat Oncol Biol Phys*, 89(1), 13–20.

Bentzen, S. M. and Thames, H. D. (1996). Tumor volume and local control probability: clinical data and radiobiological interpretations. *Int J Radiat Oncol Biol Phys*, 36(1), 247–251.

Berk, L., Berkey, B., Rich, T. et al. (2007). Randomized phase II trial of high-dose melatonin and radiation therapy for RPA class 2 patients with brain metastases (RTOG 0119). *Int J Radiat Oncol Biol Phys*, 68(3), 852–857.

Bertrand, G., Maalouf, M., Boivin, A. et al. (2014). Targeting head and neck cancer stem cells to overcome resistance to photon and carbon ion radiation. *Stem Cell Rev*, 10, 114–126.

Bese, N. S., Hendry, J., and Jeremic, B. (2007). Effects of prolongation of overall treatment time due to unplanned interruptions during radiotherapy of different tumor sites and practical methods for compensation. *Int J Radiat Oncol Biol Phys*, 68(3), 654–661.

Bhatnagar, P., Subesinghe, M., Patel, C. et al. (2013). Functional imaging for radiation treatment planning, response assessment, and adaptive therapy in head and neck cancer. *Radiographics*, 33(7), 1909–1929.

Bindra, R. S. and Glazer, P. M. (2005). Genetic instability and the tumor microenvironment: towards the concept of microenvironment-induced mutagenesis. *Mutat Res*, 569, 75–85.

Bindra, R. S., Schaffer, P. J., Meng, A. et al. (2004). Down-regulation of Rad51 and decreased homologous recombination in hypoxic cancer cells. *Mol Cell Biol*, 24, 8504–8518.

Bishop, J. A., Lewis, J. S., Rocco, J. W. et al. (2015). HPV-related squamous cell carcinoma of the head and neck: an update on testing in routine pathology practice. *Sem Diagnost Pathol*, 32, 344–351.

Bishop, J. A., Ma, X. J., Wang, H. et al. (2012). Detection of transcriptionally active high-risk HPV in patients with head and neck squamous cell carcinoma as visualized by a novel E6/E7 mRNA in situ hybridization method. *Am J Surg Pathol*, 36(12), 1874–1882.

Bittner, M. I. and Grosu, A. L. (2013). Hypoxia in head and neck tumors: characteristics and development during therapy. *Front Oncol*, 3, 233.

Bittner, M. I., Wiedenmann, N., Bucher, S. et al. (2013). Exploratory geographical analysis of hypoxic subvolumes using 18F-MISO-PET imaging in patients with head and neck cancer in the course of primary chemoradiotherapy. *Radiother Oncol*, 108, 511–516.

Blanchard, P., Baujat, B., Holostenco, V. et al. (2011). Meta-analysis of chemotherapy in head and neck cancer (MACH-NC): a comprehensive analysis by tumour site. *Radiother Oncol*, 100(1), 33–40.

Blot, W. J., McLaughlin, J. K., Winn, D. M. et al. (1988). Smoking and drinking in relation to oral and pharyngeal cancer. *Cancer Res*, 48(11), 3282–3287.

Bollineni, V. R., Kramer, G. M., Jansma, E. P. et al. (2016). A systematic review on [^{18}F]FLT-PET uptake as a measure of treatment response in cancer patients. *Eur J Cancer*, 55, 81–97.

Bonner, J. A., Harari, P. M., Giralt, J. et al. (2006). Radiotherapy plus cetuximab for squamous cell carcinoma of the head and neck. *N Engl J Med.* 354, 567–578.

Boonkitticharoen, V., Kulapaditharom, B., Leopairut, J. et al. (2008). Vascular endothelial growth factor A and proliferation marker in prediction of lymph node metastasis in oral and pharyngeal squamous cell carcinoma. *Arch Otolaryngol Head Neck Surg*, 134, 1305–1311.

Bossi, P., Resteghini, C., Paielli, N. et al. (2016). Prognostic and predictive value of EGFR in head and neck squamous cell carcinoma. *Oncotarget*, 7(45), 74362–74379.

Bourhis, J., Overgaard, J., Audry, H. et al. (2006). Hyperfractionated or accelerated radiotherapy in head and neck cancer: a meta-analysis. *Lancet*, 368(9538), 843–854.

Bourhis, J., Sire, C., Graff, P. et al. (2012). Concomitant chemoradiotherapy versus acceleration of radiotherapy with or without concomitant chemotherapy in locally advanced head and neck carcinoma (GORTEC 99-02): an open-label phase 3 randomised trial. *Lancet Oncol*, 13, 145–153.

Bozec, A., Peyrade, F., and Milano, G. (2013). Molecular targeted therapies in the management of head and neck squamous cell carcinoma: recent developments and perspectives. *Anticancer Agents Med Chem*, 13(3), 389–402.

Braunholz, D., Saki, M., Niehr, F. et al. (2016). Spheroid culture of head and neck cancer cells reveals an important role of EGFR signalling in anchorage independent survival. *PLoS One*, 11(9), e0163149.

Brenner, D. J. (1993). Dose, volume, and tumor-control predictions in radiotherapy. *Int J Radiat Oncol Biol Phys*, 26, 171–179.

Bristow, R. G. and Hill, R. P. (2008). Hypoxia and metabolism. Hypoxia, DNA repair and genetic instability. *Nat Rev Cancer*, 8, 180–192.

Brizel, D. M. (1998). Radiotherapy and concurrent chemotherapy for the treatment of locally advanced head and neck squamous cell carcinoma. *Sem Rad Oncol*, 8, 237–246.

Brizel, D. M., Albers, M. E., Fisher, S. R. et al. (1998). Hyperfractionated irradiation with or without concurrent chemotherapy for locally advanced head and neck cancer. *N Engl J Med.* 338, 1798–1804.

Brizel, D. M., Dodge, R. K., Clough, R. W. et al. (1999). Oxygenation of head and neck cancer: changes during radiotherapy and impact on treatment outcome. *Radiother Oncol*, 53, 113–117.

Brizel, D. M., Murphy, B. A., Rosenthal, D. I. et al. (2008). Phase II study of palifermin and concurrent chemoradiation in head and neck squamous cell carcinoma. *J Clin Oncol.* 26(15), 2489–2496.

Brizel, D. M., Scully, S. P., Harrelson, J. M. et al. (1996). Tumor oxygenation predicts for the likelihood of distant metastases in human soft tissue sarcoma. *Cancer Res*, 56, 941–943.

Brizel, D. M., Sibley, G. S., Prosnitz, L. R. et al. (1997). Tumor hypoxia adversely affects the prognosis of carcinoma of the head and neck. *Int J Radiat Oncol Biol Phys*, 38, 285–289.

Broders, A. C. (1920). Squamous cell cancer of the lip: a study of five hundred and thirty-seven cases. *JAMA*, 74, 656–664.

Broglie, M. A., Soltermann, A., Rohrbach, D. et al. (2013). Impact of p16, p53, smoking, and alcohol on survival in patients with oropharyngeal squamous cell carcinoma treated with primary intensity-modulated chemoradiation. *Head Neck*, 35(12), 1698–1706.

Browman, G., Hodson, I., Mackenzie, R. et al. (2001). Choosing a concomitant chemotherapy and radiotherapy regimen for squamous cell head and neck cancer: a systematic review of the published literature with subgroup analysis. *Head Neck*, 20, 579–589.

Browman, G. P., Cripps, C., Hodson, D. I. et al. (1994). Placebo-controlled randomized trial of infusional fluorouracil during standard radiotherapy in locally advanced head and neck cancer. *J Clin Oncol*, 12, 2648–2653.

Brown, J. M. (1979). Evidence for acutely hypoxic cells in mouse tumours, and a possible mechanism of reoxygenation. *Br J Radiol*, 52, 650–656.

Brown, J. M. (1999). The hypoxic cell: a target for selective cancer therapy. *Cancer Res*, 59, 5863–5870.

Brunner, T. B., Kunz-Schughart, L. A., Grosse-Gehling, P. et al. (2012). Cancer stem cells as a predictive factor in radiotherapy. *Semin Radiat Oncol*, 22(2), 151–174.

Bu, R., Purushotham, K. R., Kerr, M. et al. (1996). Alterations in the of phosphotyrosine signal transduction constituents in human parotid tumours. *Proc Soc Exp Biol Med*, 211, 257–264.

Budach, V., Stromberger, C., Poettgen, C. et al. (2015). Hyperfractionated accelerated radiation therapy (HART) of 70.6 Gy with concurrent 5-FU/Mitomycin C is superior to HART of 77.6 Gy alone in locally advanced head and neck cancer: long-term results of the ARO 95-06 randomized phase III trial. *Int J Radiat Oncol Biol Phys*, 91(5), 916–924.

Budach, W., Hehr, T., Budach, V. et al. (2006). A meta-analysis of hyperfractionated and accelerated radiotherapy and combined chemotherapy and radiotherapy regimens in unresected locally advanced squamous cell carcinoma of the head and neck. *BMC Cancer*, 6(28), 1–12.

Burtness, B., Goldwasser, M. A., Flood, W. et al. (2005). Phase III randomized trial of cisplatin plus placebo compared with cisplatin plus cetuximab in metastatic/recurrent head and neck cancer: an Eastern Cooperative Oncology Group study. *J Clin Oncol*, 23, 8646–8654.

Busch, C. J., Becker, B., Kriegs, M. et al. (2016). Similar cisplatin sensitivity of HPV-positive and -negative HNSCC cell lines. *Oncotarget*, 7(24), 35832–35842.

Busch, C. J., Kriegs, M., Laban, S. et al. (2013). HPV-positive HNSCC cell lines but not primary human fibroblasts are radiosensitized by the inhibition of Chk1. *Radiother Oncol*, 108(3), 495–499.

Busch, C. J., Kröger, M. S., Jensen, J. et al. (2017). G2-checkpoint targeting and radiosensitization of HPV/p16-positive HNSCC cells through the inhibition of Chk1 and Wee1. *Radiother Oncol*, 122(2), 260–266.

Bussink, J., Kaanders, J. H. A. M., Rijken, P. F. J. W. et al. (1999). Vascular architecture and microenvironmental parameters in human squamous cell carcinoma xenografts: effects of carbogen and nicotinamide. *Radiother Oncol*, 50(2), 173–184.

Bussink, J., van der Kogel, A. J., and Kaanders, J. H. (2008). Activation of the PI3-K/AKT pathway and implications for radioresistance mechanisms in head and neck cancer. *Lancet Oncol*, 9(3), 288–296.

Bussink, J., van Herpen, C. M., Kaanders, J. H. et al. (2010). PET-CT for response assessment and treatment adaptation in head and neck cancer. *Lancet Oncol*, 11, 661–669.

Bütof, R., Dubrovska, A., and Baumann, M. (2013). Clinical perspectives of cancer stem cell research in radiation therapy. *Radiother Oncol*, 108(3), 388–396.

Cabrera, M. C., Hollingsworth, R. E., and Hurt, E. M. (2015). Cancer stem cell plasticity and tumor hierarchy. *World J Stem Cells*, 7(1), 27–36.

Calabro-Jones, P. M., Fahey, R. C., Smoluk, G. D. et al. (1985). Alkaline phosphatase promotes radioprotection and accumulation of WR-1065 in V79-171 cells incubated in medium containing WR-2721. *Int J Radiat Biol Relat Stud Phys Chem Med*, 47(1), 23–27.

Califano, J., van der Riet, P., Westra, W. et al. (1996). Genetic progression model for head and neck cancer: implications for field cancerization. *Cancer Res*, 56(11), 2488–2492.

Campa, V. M., Gutiérrez-Lanza, R., Cerignoli, F. et al. (2008). Notch activates cell cycle reentry and progression in quiescent cardiomyocytes. *J Cell Biol*, 183, 129–141.

Campbell, N. P., Hensing, T. A., Bhayani, M. K. et al. (2016). Targeting pathways mediating resistance to anti-EGFR therapy in squamous cell carcinoma of the head and neck. *Expert Rev Anticancer Ther*, 16, 847–858.

Campos, B., Wan, F., Farhadi, M. et al. (2010). Differentiation therapy exerts antitumor effects on stem-like glioma cells. *Clin Cancer Res*, 16, 2715–2728.

Cancer Genome Atlas Network. (2015). Comprehensive genomic characterization of head and neck squamous cell carcinomas. *Nature*, 517(7536), 576–582.

Capizzi, R. L., Scheffler, B., and Scein, P. S. (1993). Amifostine-mediated protection of normal bone marrow from cytotoxic chemotherapy. *Cancer*, 72, 3495–3501.

Cardenas-Navia, L., Mace, D., Richardson, R. et al. (2008). The pervasive presence of fluctuating oxygenation in tumors. *Cancer Res*, 68, 5812–5819.

Carpenter, G. and Cohen, S. (1979). Epidermal growth factor. *Annu Rev Biochem*, 48, 193–216.

Cassatt, D. R., Fazenbaker, C. A., Kifle, G. et al. (2002). Preclinical studies on the radioprotective efficacy and pharmacokinetics of subcutaneously administered amifostine. *Semin Oncol*, 29, 2–8.

Castaldi, P., Rufini, V., Bussu, F. et al. (2012). Can "early" and "late" 18F-FDG PET–CT be used as prognostic factors for the clinical outcome of patients with locally advanced head and neck cancer treated with radio-chemotherapy? *Radiother Oncol*, 103, 63–68.

Chadwick, K. H. and Leenhouts, H. P. (1973). A molecular theory of cell survival. *Phys Med Biol*, 18(1), 78–87.

Chai, R. C., Lim, Y., Frazer, I. H. et al. (2016). A pilot study to compare the detection of HPV-16 biomarkers in salivary oral rinses with tumour p16(INK4a) expression in head and neck squamous cell carcinoma patients. *BMC Cancer*, 16, 178.

Chan, N., Ali, M., McCallum, G. P. et al. (2014). Hypoxia provokes base excision repair changes and a repair-deficient, mutator phenotype in colorectal cancer cells. *Mol Cancer Res*, 12(10), 1407–1415.

Chan, N., Koritzinsky, M., Zhao, H. et al. (2008). Chronic hypoxia decreases synthesis of homologous recombination proteins to offset chemoresistance and radioresistance. *Cancer Res*, 68, 605–614.

Chao, K. S., Bosch, W. R., Mutic, S. et al. (2001). A novel approach to overcome hypoxic tumor resistance: Cu-ATSM-guided intensity-modulated radiation therapy. *Int J Radiat Oncol Biol Phys*, 49, 1171–1182.

Chaplin, D. J., Olive, P. L., and Durand, R. E. (1987). Intermittent blood flow in a murine tumor: radiobiological effects. *Cancer Res*, 47, 597–601.

Chapman, J. D., Franko, A. J., and Sharplin, J. (1981). A marker for hypoxic cells in tumours with potential clinical applicability. *Br J Cancer*, 43, 546–550.

Chaturvedi, A., Anderson, W. F., Lortet-Tieulent, J. et al. (2013). World-wide trends in incidence rates for oral cavity and oropharyngeal cancers. *J Clin Oncol*, 31(36), 4550–4559.

Chaturvedi, A. K., Engels, E. A., Anderson, W. F. et al. (2008). Incidence trends for human papillomavirus-related and -unrelated oral squamous cell carcinomas in the United States. *J Clin Oncol*, 26(4), 612–619.

Chemotherapy With or Without Bevacizumab in Treating Patients With Recurrent or Metastatic Head and Neck Squamous Cell Carcinoma. Available from https://clinicaltrials.gov/show/NCT00588770. (Accessed on 9/11/2016.)

Chen, A. M., Bucci, M. K., Quivey, J. M. et al. (2006). Long-term outcome of patients treated by radiation therapy alone for salivary gland carcinomas. *Int J Radiat Oncol Biol Phys*, 66, 1044–1050.

Chen, A. M., Farwell, D. G., Luu, Q. et al. (2011). Prospective trial of high-dose reirradiation using daily image guidance with intensity-modulated radiotherapy for recurrent and second primary head-and-neck cancer. *Int J Radiat Oncol Biol Phys*, 80, 669–676.

Chen, A. M., Phillips, T. L., and Lee, N. Y. (2011). Practical considerations in the re-irradiation of recurrent and second primary head-and-neck cancer. *Int J Radiat Oncol Biol Phys*, 81, 1211–1219.

Chen, J., Zhou, J., Lu, J. et al. (2014). Significance of CD44 expression in head and neck cancer: a systemic review and meta-analysis. *BMC Cancer*, 14, 15.

Chen, L., Zhang, Z., Kolb, H. C. et al. (2012). ^{18}F-HX4 hypoxia imaging with PET/CT in head and neck cancer: a comparison with ^{18}F-FMISO. *Nucl Med Commun*, 33(10), 1096–1102.

Chen, N. X., Chen, L., Wang, J. L. et al. (2016). A clinical study of multimodal treatment for orbital organ preservation in locally advanced squamous cell carcinoma of the nasal cavity and paranasal sinus. *Jpn J Clin Oncol*, 46(8), 727–734.

Chen, Y. S., Wu, M. J., Huang, C. Y. et al. (2011). CD133/Src axis mediates tumor initiating property and epithelial-mesenchymal transition of head and neck cancer. *PloS One*, 6(11), e28053.

Cheng, J., Tian, L., Ma, J. et al. (2015). Dying tumor cells stimulate proliferation of living tumor cells via caspase-dependent protein kinase Cδ activation in pancreatic ductal adenocarcinoma. *Mol Oncol*, 9(1), 105–114.

Cheung, T. H. and Rando, T. A. (2013). Molecular regulation of stem cell quiescence. *Nat Rev Mol Cell Biol*, 14(6), 329–340.

Chiang, Y. C., Teng, S. C., Su, Y. N. et al. (2003). c-MYC directly regulates the transcription of NBS1 gene involved in DNA double-strand break repair. *J Biol Chem*, 278, 19286–19291.

Chikamatsu, K., Ishii, H., Murata, T. et al. (2013). Alteration of cancer stem cell-like phenotype by histone deacetylase inhibitors in squamous cell carcinoma of the head and neck. *Cancer Sci*, 104, 1468–1475.

Choe, K. S., Haraf, D. J., Solanki, A. et al. (2011). Prior chemoradiotherapy adversely impacts outcomes of recurrent and second primary head and neck cancer treated with concurrent chemotherapy and reirradiation. *Cancer*, 117, 4671–4678.

Choong, N. W., Kozloff, M., Taber, D. et al. (2010). Phase II study of sunitinib malate in head and neck squamous cell carcinoma. *Investigational New Drugs*, 28(5), 677–683.

Christopoulos, A., Ahn, S. M., Klein, J. D. et al. (2011). Biology of vascular endothelial growth factor and its receptors in head and neck cancer: beyond angiogenesis. *Head Neck*, 33(8), 1220–1229.

Chu, F. C., Conrad, J. T., Bane, H. N. et al. (1960). Quantitative and qualitative evaluation of skin erythema I. Technic of measurement and description of the reaction. *Radiology*, 75, 406–410.

Chua, M. L., Wee, J. T., Hui, E. P. et al. (2016). Nasopharyngeal carcinoma. *Lancet*, 387(10022), 1012–1024.

Chung, C. H., Dignam, J. J., Hammond, M. E. et al. (2011). Glioma-associated oncogene family zinc finger 1 expression and metastasis in patients with head and neck squamous cell carcinoma treated with radiation therapy (RTOG 9003). *J Clin Oncol*, 29, 1326–1334.

Ciardello, F. and Tortora, G. (2001). A novel approach in the treatment of cancer: targeting the epidermal growth factor receptor. *Clin Cancer Res*, 7, 2958–2970.

Cicalese, A., Bonizzi, G., Pasi, C. E. et al. (2009). The tumor suppressor p53 regulates polarity of self-renewing divisions in mammary stem cells. *Cell*, 138(6), 1083–1095.

Citrin, D., Cotrim, A. P., Hyodo, F. et al. (2010). Radioprotectors and mitigators of radiation-induced normal tissue injury. *Oncologist*, 15(4), 360–371.

Clavel, S., Nguyen, D. H. A., Fortin, B. et al. (2012). Simultaneous integrated boost using intensity-modulated radiotherapy compared with conventional radiotherapy in patients treated with concurrent carboplatin and 5-fluorouracil for locally advanced oropharyngeal carcinoma. *Int J Radiat Oncol Biol Phys*, 82, 582–589.

Clay, M. R., Tabor, M., Owen, J. H. et al. (2010). Single-marker identification of head and neck squamous cell carcinoma cancer stem cells with aldehyde dehydrogenase. *Head Neck*, 32(9), 1195–1201.

Cloos, J., Nieuwenhuis, E. J., Boomsma, D. I. et al. (1996). Genetic susceptibility to head and neck squamous cell carcinoma. *J Natl Cancer Inst*, 88(8), 530–535.

Codeca, C., Ferrari, D., Bertuzzi, C. et al. (2012). Angiogenesis in head and neck cancer: a review of the literature. *J Oncol*, 2012, 358472.

Cohen, E. E., Davis, D. W., Karrison, T. G. et al. (2009). Erlotinib and bevacizumab in patients with recurrent or metastatic squamous-cell carcinoma of the head and neck: a phase I/II study. *Lancet Oncol*, 10(3), 247–257.

Cohen, E. E., Lingen, M. W., and Vokes, E. E. (2004). The expanding role of systemic therapy in head and neck cancer. *J Clin Oncol*, 22(9), 1743–1752.

Collingridge, D. R., Young, W. K., Vojnovic, B. et al. (1997). Measurement of tumor oxygenation: a comparison between polarographic needle electrodes and a time-resolved luminescence-based optical sensor. *Rad Res*, 147, 329–334.

Combes, J. D. and Franceschi, S. (2014). Role of human papillomavirus in non-oropharyngeal head and neck cancers. *Oral Oncol*, 50, 370–379.

Cooper, J. S., Fu, K., Marks, J. et al. (1995). Late effects of radiation therapy in the head and neck region. *Int J Radiat Oncol Biol Phys*, 31, 1141–1164.

Coppes, R. P., Zeilstra, L. J. W., Kampinga, H. H. et al. (2001). Early to late sparing of radiation damage to the parotid gland by adrenergic and muscarinic receptor agonists. *Br J Cancer*, 85(7), 1055–1063.

Corry, J., Peters, L. J., and Rischin, D. (2010). Optimising the therapeutic ratio in head and neck cancer. *Lancet Oncol*. 11(3), 287–291.

Corvo, R. (2007). Evidence–based radiation oncology in head and neck squamous cell carcinoma. *Radiother Oncol*, 85, 156–170.

Cotrim, A. P., Hyodo, F., Matsumoto, K. et al. (2007). Differential radiation protection of salivary glands versus tumor by Tempol with accompanying tissue assessment of Tempol by magnetic resonance imaging. *Clin Cancer Res*, 13(16), 4928–4933.

Cottrill, C. P., Bishop, K., Walton, M. I. et al. (1998). Pilot study of nimorazole as a hypoxic-cell sensitizer with the 'CHART' regimen in head and neck cancer. *Int J Radiat Oncol Biol Phys*, 42(4), 807–810.

Coutard, H. (1932). Roentgen therapy of epitheliomas of the tonsillar region, hypopharynx and larynx from 1920 to 1926. *Am J Roentgenol*, 27, 313–331.

Cox, J. D., Stetz, J., and Pajak, T. F. (1995). Toxicity criteria of the Radiation Therapy Oncology Group (RTOG) and the European Organization for Research and Treatment of Cancer (EORTC). *Int J Radiat Oncol Biol Phys*, 31, 1341–1346.

Crabtree, H. G. and Cramer, W. (1933). The action of radium on cancer cells. II.—some factors determining the susceptibility of cancer cells to radium. *Proc R Soc London*, 113, 238–250.

Crile, G. W. (1906). On the surgical treatment of cancer of the head and neck. *Trans South Surg Gynecol Assoc*, 18, 108–127.

Cummings, B., Keane, T., Pintilie, M. et al. (2007). Five year results of a randomized trial comparing hyperfractionated to conventional radiotherapy over four weeks in locally advanced head and neck cancer. *Radiother Oncol*, 85, 7–16.

Curtis, S. B. (1986). Lethal and potentially lethal lesions induced by radiation—a unified repair model. *Radiat Res*, 106(2), 252–270.

Dale, R. and Jones, B. (Eds.). Radiobiological Modelling in Radiation Oncology. British Institute of Radiology; 2007.

Dale, R. G., Hendry, J. H., Jones, B. et al. (2002). Practical methods for compensating for missed treatment days in radiotherapy, with particular reference to head and neck schedules. *Clin Oncol*, 14(5), 382–393.

Dale, R. G. and Jones, B. (1999). The assessment of RBE effects using the concept of biologically effective dose. *Int J Radiat Oncol Biol Phys*, 43(3), 639–645.

Das, L. C., Karrison, T. G., Witt, M. E. et al. (2015). Comparison of outcomes of locoregionally advanced oropharyngeal and non-oropharyngeal squamous cell carcinoma over two decades. *Ann Oncol*, 26(1), 198–205.

Dasari, S. and Tchounwou, P. B. (2014). Cisplatin in cancer therapy: molecular mechanisms of action. *Eur J Pharmacol*, 740, 364–378.

Dassonville, O., Formento, J. L., Francoual, M. et al. (1993). Expression of epidermal growth factor receptor and survival in upper aerodigestive tract cancer. *J Clin Oncol*, 11, 1873–1878.

Dasu, A. Modelling the impact of two forms of hypoxia on novel radiotherapy approaches. Ph.D. thesis, Umeå University, 2000.

Dasu, A. and Denekamp, J. (1998). New insights into factors influencing the clinically relevant oxygen enhancement ratio. *Radiother Oncol*, 46, 269–277.

Dasu, A. and Denekamp, J. (2003). The impact of tissue microenvironment on treatment simulation. *Adv Exp Med Biol*, 510, 63–67.

Dasu, A. and Toma-Dasu, I. (2013). Impact of variable RBE on proton fractionation. *Med Phys*, 40(1), 011705.

Dasu, A. and Toma-Dasu, I. (2008). What is the clinically relevant relative biologic effectiveness? A warning for fractionated treatments with high linear energy transfer radiation. *Int J Radiat Oncol Biol Phys*, 70, 867–874.

Datta, N. R., Rogers, S., Ordóñez, S. G. et al. (2016). Hyperthermia and radiotherapy in the management of head and neck cancers: a systematic review and meta-analysis *Int J Hyperthermia*, 32(1), 31–40.

Davis, A. and Tannock, I. (2000). Repopulation of tumour cells between cycles of chemotherapy: a neglected factor, *Lancet Oncol*, 1(2), 86–93.

de Bree, R. and Leemans, C. R. (2010). Recent advances in surgery for head and neck cancer. *Curr Opin Oncol*, 22(3), 186–193.

De Crevoisier, R., Bourhis, J., Domenge, C. et al. (1998). Full-dose reirradiation for unresectable head and neck carcinoma: experience at the Gustave-Roussy Institute in a series of 169 patients. *J Clin Oncol*, 16, 3556–3562.

De Felice, F., de Vincentiis, M., Valentini, V. et al. (2017). Follow-up program in head and neck cancer. *Crit Rev Oncol Hematol*, 113, 151–155.

de Jong, M. C., Pramana, J., van der Wal, J. E. et al. (2010). CD44 expression predicts local recurrence after radiotherapy in larynx cancer. *Clin Cancer Res*, 16, 5329–5338.

De Meulenaere, A., Vermassen, T., Aspeslagh, S. et al. (2017). Tumor PD-L1 status and CD8+ tumor-infiltrating T cells: markers of improved prognosis in oropharyngeal cancer. *Oncotarget*, 8(46), 80443–80452. doi: 10.18632/oncotarget.19045.

Dehdashti, F., Laforest, R., Gao, F. et al. (2013). Assessment of cellular proliferation in tumors by PET using 18F-ISO-1. *J Nucl Med*, 54(3), 350–357.

Denekamp, J. and Dasu, A. (1999). Inducible repair and the two forms of tumour hypoxia—time for a paradigm shift. *Acta Oncol*, 38, 903–918.

Denekamp, J., Dasu, A., and Waites, A. (1998). Vasculature and microenvironmental gradients: the missing links in novel approaches to cancer therapy? *Advances in Enzyme Regulation*, 38(1), 281–299.

Deng, W., Lin, B. Y., Jin, G. et al. (2004). Cyclin/CDK regulates the nucleocytoplasmic localization of the human papillomavirus E1 DNA helicase. *J Virol*, 78(24), 13954–13965.

Denham, J., Walker, Q. J., Lamb, D. S. et al. (1996). Mucosal regeneration during radiotherapy. *Radiother Oncol*, 41, 109–118.

Denham, J. W. and Abbott, R. L. (1991). Concurrent cisplatin, infusional fluorouracil, and conventionally fractionated radiation therapy in head and neck cancer: dose-limiting mucosal toxicity. *J Clin Oncol*, 9, 458–463.

Denham, J. W., Walker, Q. J., Lamb, D. S. et al. (1996). Mucosal regeneration during radiotherapy: Trans Tasman Radiation Oncology Group (TROG). *Radiother Oncol*, 41, 109–118.

Determination of Cetuximab Versus Cisplatin Early and Late Toxicity Events in HPV+ OPSCC (De-ESCALaTE). Available from https://clinicaltrials.gov/ct2/show/study/NCT01874171. (Accessed on 16/01/2017.)

Dhar, S. and Lippard, S. J. (2009). Mitaplatin, a potent fusion of cisplatin and the orphan drug dichloroacetate. *Proc Nat Acad Sci USA*, 106(52), 22199–22204.

Dirix, P., Vandecaveye, V., De Keyzer, F. et al. (2009). Dose painting in radiotherapy for head and neck squamous cell carcinoma: value of repeated functional imaging with (18)F-FDG PET, (18)F-fluoromisonidazole PET, diffusion-weighted MRI, and dynamic contrast-enhanced MRI. *J Nucl Med*, 50(7), 1020–1027.

Dische, D., Saunders, M., Barrett, A. et al. (1997). A randomised multicentre trial of CHART versus conventional radiotherapy in head and neck cancer. *Radiother Oncol*, 44, 123–136.

Dische, S., Warburton, M. F., Jones, D. et al. (1989). The recording of morbidity related to radiotherapy. *Radiother Oncol*, 16, 103–108.

Domenge, C., Hill, C., Lefebvre, J. et al. (2000). Randomized trial of neoadjuvant chemotherapy in oropharyngeal carcinoma. *Br J Cancer*, 83(12), 1594–1598.

Dörr, W. (1997). Three A's of repopulation during fractionated irradiation of squamous epithelia: asymmetry loss, acceleration of stem-cell divisions and abortive divisions. *Int J Radiat Biol*, 72, 635–643.

Dörr, W. (2003). Modulation of repopulation processes in oral mucosa: experimental results. *Int J Radiat Biol*, 79(7), 531–537.

Dörr, W. Time factors in normal-tissue responses to irradiation. In: Van der Koegel A, Joiner M (Eds.). Basic Clinical Radiobiology, 4th edition. Oxford University Press; 2009. pp. 149–157.

Dörr, W., Hamilton, C. S., Boyd, T. et al. (2002). Radiation-induced changes in cellularity and proliferation in human oral mucosa. *Int J Radiat Oncol Biol Phys*, 52, 911–917.

Dörr, W. and Hendry, J. H. (2001). Consequential late effects in normal tissues. *Radiother Oncol*, 61, 223–231.

Dorsey, K. and Agulnik, M. (2013). Promising new molecular targeted therapies in head and neck cancer. *Drugs*, 73(4), 315–325.

Douglas, B. G. and Fowler, J. F. (1976). The effect of multiple small doses of x rays on skin reactions in the mouse and a basic interpretation. *Radiat Res*, 66, 401–420.

Douglas, J. G., Koh, W. J., Austin-Seymour, M. et al. (2003). Treatment of salivary gland neoplasms with fast neutron radiotherapy. *Arch Otolaryngol Head Neck Surg*, 129(9), 944–948.

Douglas, J. G., Laramore, G. E., Austin-Seymour, M. et al. (1996). Neutron radiotherapy for adenoid cystic carcinoma of minor salivary glands. *Int J Radiat Oncol Biol Phys*, 36, 87–93.

Dubben, H. H., Thames, H. D., and Beck–Bornholdt, H. P. (1998). Tumor volume: a basic and specific response predictor in radiotherapy. *Radiother Oncol*, 47, 167–174.

Dunst, J., Stadler, P., Becker, A. et al. (2003). Tumor volume and tumor hypoxia in head and neck cancers. The amount of the hypoxic volume is important. *Strahlenther Onkol*, 179, 521–526.

Eisenhauer, E. A., Therasse, P., Bogaerts, J. et al. (2009). New response evaluation criteria in solid tumours: revised RECIST guideline (version 1.1). *Eur J Cancer*, 45, 228–247.

Ekshyyan, O., Mills, G. M., Lian, T. et al. (2010). Pharmacodynamic evaluation of temsirolimus in patients with newly diagnosed advanced-stage head and neck squamous cell carcinoma. *Head Neck*, 32(12), 1619–1628.

Elkind, M. M. and Sutton, H. (1960). Radiation response of mammalian cells grown in culture: I. Repair of x-ray damage in surviving Chinese hamster cells. *Radiat Res*, 13, 556–593.

Ellis, F. (1969). Dose, time and fractionation: a clinical hypothesis. *Clin Radiol*, 20, 1–7.

Elser, C., Siu, L. L., Winquist, E. et al. (2007). Phase II trial of sorafenib in patients with recurrent or metastatic squamous cell carcinoma of the head and neck or nasopharyngeal carcinoma. *J Clin Oncol*, 25, 3766–3773.

Emami, B., Bignardi, M., Spector, G. J. et al. (1987). Reirradiation of recurrent head and neck cancers. *Laryngoscope*, 97, 85–88.

Emami, B., Lyman, J., Brown, A. et al. (1991). Tolerance of normal tissue to therapeutic radiation. *Int J Radiat Oncol Biol Phys*, 21, 109–122.

Epstein, R. J. (2007). VEGF signaling inhibitors: more pro-apoptotic than anti-angiogenic. *Cancer Metastasis Rev*, 26, 443–452.

Eschmann, S. M., Paulsen, F., Bedeshem, C. et al. (2007). Hypoxia-imaging with (18)F-Misonidazole and PET: changes of kinetics during radiotherapy of head-and-neck cancer. *Radiother Oncol*, 83, 406–410.

Eschmann, S. M., Paulsen, F., Reimold, M. et al. (2005). Prognostic impact of hypoxia imaging with 18F-misonidazole PET in non-small cell lung cancer and head and neck cancer before radiotherapy. *J Nucl Med*, 46, 253–260.

Espinosa, M., Martinez, M., Aquilar, J. L. et al. (2005). Oxaliplatin activity in head and neck cancer cell lines. *Cancer Chemother Pharmacol*, 55(3), 301–305.

Evans, S. M., Hahn, S., Pook, D. R. et al. (2000). Detection of hypoxia in human squamous cell carcinoma by EF5 binding. *Cancer Res*, 60, 2018–2024.

Evans, S. M., Hahn, S. M., Magarelli, D. P. et al. (2001). Hypoxia in human intraperitoneal and extremity sarcomas. *Int J Radiat Oncol Biol Phys*, 49, 587–596.

Evans, S. M., Joiner, B., Jenkins, W. T. et al. (1995). Identification of hypoxia in cells and tissues of epigastric 9L rat glioma using EF5 [2-(2-nitro-1H-imidazol-1-yl)-N-(2,2,3,3,3-pentafluoropropyl) acetamide]. *Br J Cancer*, 72, 875–882.

Fan, K. F., Hopper, C., Speight, P. M. et al. (1996). Photodynamic therapy using 5-aminolevulinic acid for premalignant and malignant lesions of the oral cavity. *Cancer*, 78, 1374–1383.

Fearon, E. R. and Vogelstein, B. A. (1990). A genetic model for colorectal tumorigenesis. *Cell*, 61, 759–767.

Ferguson, B. J., Hudson, W. R., and Farmer, J. C. (1987). Hyperbaric oxygen therapy for laryngeal radionecrosis. *Ann Otol Rhinol Laryngol*, 96(1), 1–6.

Ferlay, J., Soerjomataram, I., Ervik, M. et al. GLOBOCAN 2012 v1.0, Cancer Incidence and Mortality Worldwide: IARC CancerBase No. 11, 2013, Lyon, France: International Agency for Research on Cancer. Available from http://globocan.iarc.fr. (Accessed on 3/03/2017.)

Ferrara, N. (2005). VEGF as a therapeutic target in cancer. *Oncology*, 69(Suppl 3), 11–16.

Ferrara, N., Gerber, H. P., and LeCouter, J. (2003). The biology of VEGF and its receptors. *Nat Med*, 9, 669–676.

Ferrara, N., Hillan, K. J., Gerber, H. P. et al. (2004). Discovery and development of bevacizumab, an anti-VEGF antibody for treating cancer. *Nat Rev Drug Discov*, 3(5), 391–400.

Finnegan, V., Parsons, J. T., Greene, B. D. et al. (2009). Neoadjuvant chemotherapy followed by concurrent hyperfractionated radiation therapy and sensitizing chemotherapy for locally advanced (T3-T4) oropharyngeal squamous cell carcinoma. *Head Neck*, 31, 167–174.

Fisch, U. (1983). The infratemporal fossa approach for nasopharyngeal tumors. *Laryngoscope*, 93, 36–44.

Fletcher, G. H. (1986). History of irradiation in squamous cell carcinomas of the larynx and hypopharynx. *Int J Radiat Oncol Biol Phys*, 12(11), 2019–2024.

Flynn, R. T., Bowen, S. R., Bentzen, S. M. et al. (2008). Intensity-modulated x-ray (IMXT) versus proton (IMPT) therapy for theragnostic hypoxia-based dose painting. *Phys Med Biol*, 53, 4153–4167.

Ford, S. E., Brandwein-Gensler, M., Carroll, W. R. et al. (2014). Transoral robotic versus open surgical approaches to oropharyngeal squamous cell carcinoma by human papillomavirus status. *Otolaryngol Head Neck Surg*, 151(4), 606–611.

Fountzilas, G., Fragkoulidi, A., Kalogera-Fountzila, A. et al. (2010). A phase II study of sunitinib in patients with recurrent and/or metastatic non-nasopharyngeal head and neck cancer. *Cancer Chemother Pharmacol*, 65, 649–660.

Fowler, J. F. (1989). The linear-quadratic formula and progress in fractionated radiotherapy: a review. *Br J Radiol*, 62, 679–694.

Fowler, J. F. (2006). Development of radiobiology for oncology—a personal view. *Phys Med Biol*, 51, R263–R286.

Fowler, J. F. (2008). Optimum overall times II: extended modelling for head and neck radiotherapy. *Clin Oncol*, 20, 113–126.

Fowler, J. F. (2010). Years of biologically effective dose. *Br J Radiol*, 83, 554–568.

Fowler, J. F. and Stern, B. E. (1960). Dose rate effects: some theoretical and practical considerations. *Br J Radiol*, 33, 389–395.

Fowler, J. F. and Stern, B. E. (1963). Dose-time relationships in radiotherapy and the validity of cell survival curve models. *Br J Radiol*, 36, 163–173

Fowler, J. F., Dasu, A., Toma-Dasu, I. Optimum Overall Treatment Time in Radiation Oncology. Madison: Medical Physics Publishing; 2015.

Fowler, J. F., Harari, P. M., Leborgne, F. et al. (2003). Acute radiation reactions in oral and pharyngeal mucosa: tolerable levels in altered fractionation schedules. *Radiother Oncol*, 69, 161–168.

Franceschi, S., Levi, F., Dal Maso, L. et al. (2000). Cessation of alcohol drinking and risk of cancer of the oral cavity and pharynx. *Int J Cancer*, 85(6), 787–790.

Franceschi, S., Levi, F., La Vecchia, C. et al. (1999). Comparison of the effect of smoking an alcohol drinking between oral and oropharyngeal cancer. *Int J Cancer*, 83(1), 1–4.

Franceschi, S., Talamini, R., Barra, S. et al. (1990). Smoking and drinking in relation to cancers of the oral cavity, pharynx, larynx and esophagus in northern Italy. *Cancer Res*, 50(20), 6502–6507.

Franzén, L., Henriksson, R., Littbrand, B. et al. (1995). Effects of sucralfate on mucositis during and following radiotherapy of malignancies in the head and neck region. A double-blind placebo-controlled study. *Acta Oncol*, 34, 219–223.

Friedman, J. M., Stavas, M. J., and Cmelak, A. J. (2014). Clinical and scientific impact of human papillomavirus on head and neck cancer. *World J Clin Oncol*, 5(4), 781–791.

Frisch, M. and Biggar, R. J. (1999). Aetiological parallel between tonsillar and anogenital squamous-cell carcinomas. *Lancet*, 354(9188), 1442–1443.

Fu, K. K. (1998). Combined radiotherapy and chemotherapy for nasopharyngeal carcinoma. *Sem Rad Oncol*, 8, 247–253.

Fu, K. K., Pajak, T. F., Trotti, A. et al. (2000). A Radiation Therapy Oncology Group (RTOG) phase III randomized study to compare hyperfractionation and two variants of accelerated fractionation to standard fractionation radiotherapy for head and neck squamous cell carcinomas; first report of RTOG 90–03. *Int J Radiat Oncol Biol Phys*, 48, 7–16.

Fujita, T., Doihara, H., Washio, K. et al. (2006). Proteasome inhibitor bortezomib increases PTEN expression and enhances trastuzumab-induced growth inhibition in trastuzumab-resistant cells. *Anticancer Drugs*, 17, 455–462.

Furusawa, Y., Fukutsu, K., Aoki, M. et al. (2000). Inactivation of aerobic and hypoxic cells from three different cell lines by accelerated 3He-, 12C- and 20Ne-Ion Beams. *Rad Res*, 154, 485–496.

Fury, M. G., Lee, N. Y., Sherman, E. et al. (2012). A phase 2 study of bevacizumab with cisplatin plus intensity-modulated radiation therapy for stage III/IVB head and neck squamous cell cancer. *Cancer*, 118(20), 5008–5014.

Fury, M. G., Lee, N. Y., Sherman, E. et al. (2013). A phase I study of everolismus + weekly cisplatin + intensity modulated radiation therapy in head-and-neck cancer. *Int J Radiat Oncol Biol Phys*, 87, 479–486.

Fury, M. G., Xiao, H., Sherman, E. J. et al. (2016). Phase II trial of bevacizumab + cetuximab + cisplatin with concurrent intensity-modulated radiation therapy for patients with stage III/IVB head and neck squamous cell carcinoma. *Head Neck*, 38, E566–E570.

Gan, G. N., Eagles, J., Keysar, S. B. et al. (2014). Hedgehog signaling drives radioresistance and stroma-driven tumor repopulation in head and neck squamous cancers. *Cancer Res*, 74(23), 7024–7036.

Geissler, C., Hambek, M., Leinung, M. et al. (2012). The challenge of tumor heterogeneity-different phenotypes of cancer stem cells in a head and neck squamous cell carcinoma xenograft mouse model. *In Vivo*, 26(4), 593–598.

Gerweck, L. E., Koutcher, J., and Zaidi, S. T. (1995). Energy status parameters, hypoxia fraction and radiocurability across tumor types. *Acta Oncol*, 34, 335–338.

Gerweck, L. E., Seneviratne, T., and Gerweck, K. K. (1993). Energy status and radiobiological hypoxia at specified oxygen concentrations. *Radiat Res*, 135, 69–74.

Ghadjar, P., Simcock, M., Studer, G. et al. (2012). Concomitant cisplatin and hyperfractionated radiotherapy in locally advanced head and neck cancer: 10-year follow-up of a randomized phase III trial (SAKK 10/94). *Int J Radiat Oncol Biol Phys*, 82(2), 524–531.

Gibson, M., Kies, M., Kim, S. et al. (2009). Cetuximab (C) and bevacizumab (B) in patients with recurrent or metastatic head and neck squamous cell carcinoma: an updated report. *J Clin Oncol (Meeting Abstracts)*, 27, 6049.

Gillison, M. L., Broutian, T., Pickard, R. K. et al. (2012). Prevalence of oral HPV infection in the United States, 2009-2010. *JAMA*, 307(7), 693–703.

Gillison, M. L., Castellsague, X., Chaturvedi, A. et al. (2014). EUROGIN roadmap: comparative epidemiology of HPV infection and associated cancers of the head and neck and cervix. *Int J Cancer*, 134(3), 497–507.

Gillison, M. L., Koch, W. M., Capone, R. B. et al. (2000). Evidence for a causal association between human papillomavirus and a subset of head and neck cancers. *J Natl Cancer Inst*, 92(9), 709–720.

Giri, U., Ashorn, C. L., Ramdas, L. et al. (2006). Molecular signatures associated with clinical outcome in patients with high risk head and neck squamous cell carcinoma treated by surgery and radiation. *Int J Radiat Oncol Biol Phys*, 64, 670–677.

Go, S. R. and Adjei, A. A. (1999). Review of the comparative pharmacology and clinical activity of cisplatin and carboplatin. *J Clin Oncol*, 17(1), 409–422.

Goldstein, B. Y., Chang, S. C., Hashibe M. et al. (2010). Alcohol consumption and cancers of the oral cavity and pharynx from 1988 to 2009, an update. *Eur J Cancer Prev*, 19(6), 431–465.

Gómez-López, S., Lerner, R. G., and Petritsch, C. (2014). Asymmetric cell division of stem and progenitor cells during homeostasis and cancer. *Cell Molec Life Sciences*, 71(4), 575–597.

Goodwin, W. J. Jr. (2000). Salvage surgery for patients with recurrent squamous cell carcinoma of the upper aerodigestive tract: when do the ends justify the means? *Laryngoscope*, 110(Suppl 93), 1–18.

Gorre, M. E., Mohammed, M., Ellwood, K. et al. (2001). Clinical resistance to STI-571 cancer therapy caused by BCR-ABL gene mutation or amplification. *Science*, 293, 876–880.

Grant, W. E., Hopper, C., Speight, P. M. et al. (1993). Photodynamic therapy of malignant and premalignant lesions in patients with 'field cancerization' of the oral cavity. *J Laryngol Otol*, 107, 1140–1145.

Gray, L. H. (1944). Dosage-rate in radiotherapy. *Br J Radiol*, 17, 327–335.

Gray, L. H., Conger, A. D., Ebert, M. et al. (1953). The concentration of oxygen dissolved in tissues at the time of irradiation as a factor in radiotherapy. *Br J Radiology*, 26, 638–648.

Gregoire, V. and Chiti, A. (2011). Molecular imaging in radiotherapy planning for head and neck tumors. *J Nucl Med*, 52, 331–334.

Grégoire, V., Langendijk, J. A., and Nuyts, S. (2015). Advances in radiotherapy for head and neck cancer. *J Clin Oncol*, 33(29), 3277–3284.

Gregoire, V., Lefebvre, J. L., Licitra, L. et al. (2010). Squamous cell carcinoma of the head and neck: EHNS-ESMO-ESTRO Clinical Practice Guidelines for diagnosis, treatment and follow-up. *Ann Oncol*, 21(Suppl. 5), 184–186.

Greven, K. M., Williams, D. W., Keyes, J. W. Jr et al. (1994). Positron emission tomography of patients with head and neck carcinoma before and after high dose irradiation. *Cancer*, 74, 1355–1359.

Grosu, A. L., Souvatzoglou, M., Roper, B. et al. (2007). Hypoxia imaging with FAZA-PET and theoretical considerations with regard to dose painting for individualization of radiotherapy in patients with head and neck cancer. *Int J Radiat Oncol Biol Phys*, 69, 541–551.

Gu, J., Zhu, S., Li, X. et al. (2014). Effect of amifostine in head and neck cancer patients treated with radiotherapy: a systematic review and meta-analysis based on randomized controlled trials. *PLoS One*, 9(5):e95968.

Haffty, B. G., Hurley, R., and Peters, L. J. (1999). Radiation therapy with hyperbaric oxygen at 4 atmospheres pressure in the management of squamous cell carcinoma of the head and neck: results of a randomized clinical trial. *Cancer J Sci Am*, 5(6), 341–347.

Hagemann, R. F., Sigdestad, C. P., and Lesher, S. (1972). Intestinal crypt survival and total and per crypt levels of proliferative cellularity following irradiation: role of crypt cellularity. *Radiat Res*, 50(3), 583–591.

Hahn, C. E. (1980). Techniques for measuring the partial pressures of gases in the blood. Part I—in vitro measurements. *Journal of Physics E: Scientific Instruments*, 13, 470–482.

Hahn, R. (1904). *Fortscnr Geb Rontgenstr Nuklearmed*, 8, 120–121.

Hall, E. J. Radiobiology for the Radiologist, 5th edition. Philadelphia: Lippincott Williams & Wilkins; 2000.

Hall, E. J. (1972). The effect of hypoxia on the repair of sublethal radiation damage in cultured mammalian cells. *Radiat Res*, 49, 405–415.

Hall, E. J. and Giaccia, A. Radiobiology for the Radiologist, 7th edition. Philadelphia: Lippincott Williams & Wilkins; 2012.

Hall, E. J., Bedford, J. S., and Oliver, R. (1966). Extreme hypoxia; its effect on the survival of mammalian cells irradiated at high and low dose-rates. *Br J Radiol*, 39, 302–307.

Hammoudi, K., Pinlong, E., Kim, S. et al. (2015). Transoral robotic surgery versus conventional surgery in treatment for squamous cell carcinoma of the upper aerodigestive tract. *Head Neck*, 37(9), 1304–1309.

Han, D., Huang, Z. G., Zhang, W. et al. (2003). Effectiveness of recombinant adenovirus p53 injection on laryngeal cancer: phase I clinical trial and follow up. *Chinese J Cancer Res*, 83, 2029–2032.

Hanahan, D. and Weinberg, R. A. (2000). The hallmarks of cancer. *Cell*, 100(1), 57–70.

Hanahan, D. and Weinberg, R. A. (2011). Hallmarks of cancer: the next generation. *Cell*, 144(5), 646–674.

Hansen, O., Grau, C., Bentzen, S. et al. (1996). Repopulation in the SCCVII squamous cell carcinoma assessed by an in vivo-in vitro excision assay. *Radiother Oncol*, 39, 137–144.

Hansson, B. G., Rosenquist, K., Antonsson, A. et al. (2005). Strong association between infection with human papillomavirus an oral and oropharyngeal squamous cell carcinoma: a population-based-case-control study in southern Sweden. *Acta Otolaryngol*, 125(12):1337–1344.

Haraf, D. J., Weichselbaum, R. R., and Vokes, E. E. (1996). Reirradiation with concomitant chemotherapy of unresectable recurrent head and neck cancer: a potentially curable disease. *Ann Oncol*, 7(9), 913–918.

Harari, P. M. and Huang, S. M. (2001). Head and neck cancer as a clinical model for molecular targeting of therapy: combining EGFR blockade with radiation. *Int J Radiat Oncol Biol Phys*, 49, 427–433.

Hardisson, D. (2003). Molecular pathogenesis of head and neck squamous cell carcinoma. *Eur Arch Otorhinolaryngol*, 260, 502–508.

Harper, L., Piper, K., Common, J. et al. (2007). Stem cell patterns in cell lines derived from head and neck squamous cell carcinoma. *J Oral Pathol Med*, 36(10), 594–603.

Harrington, K. J., Ferris, R. L., Blumenschein, G. Jr et al. (2017). Nivolumab versus standard, single-agent therapy of investigator's choice in recurrent or metastatic squamous cell carcinoma of the head and neck (CheckMate 141): health-related quality-of-life results from a randomised, phase 3 trial. *Lancet Oncol*, 18(8), 1104–1115.

Harrington, K. J., Temam, S., D'Cruz, A. et al. (2015). Postoperative adjuvant lapatinib and concurrent chemoradiotherapy followed by maintenance lapatinib monotherapy in high-risk patients with resected squamous cell carcinoma of the head and neck: a phase iii, randomized, double-blind, placebo-controlled study. *J Clin Oncol*, 33(35), 4202–4209.

Harris, B. N., Biron, V. L., Donald, P. et al. (2015). Primary surgery vs chemoradiation treatment of advanced-stage hypopharyngeal squamous cell carcinoma. *JAMA Otolaryngol Head Neck Surg*, 141(7), 636–640.

Hasina, R., Whipple, M. E., Martin, L. E. et al. (2008). Angiogenic heterogeneity in head and neck squamous cell carcinoma: biological and therapeutic implications. *Lab Invest*, 88(4), 342–353.

Hatfield, S. and Ruohola-Baker, H. (2008). microRNA and stem cell function. *Cell Tissue Res*, 331(1), 57–66.

Hemminki, K., Dong, C., and Frisch, M. (2000). Tonsillar and other upper aerodigestive tract cancers among cervical cancer patients and their husbands. *Eur J Cancer Prev*, 9(6), 433–437.

Hendry, J. H. and Thames, H. D. (1986). The tissue-rescuing unit. *Br J Radiol*, 59(702), 628–630.

Hendry, J. H., Bentzen, S. M., Dale, R. G. et al. (1996). A modelled comparison of the effects of using different ways to compensate for missed treatment days in radiotherapy. *Clin Oncol*, 8, 297–307.

Henk, J. M. (1986). Late results of a trial of hyperbaric oxygen and radiotherapy in head and neck cancer: a rationale for hypoxic cell sensitizers. *Int J Radiat Oncol Biol Phys*, 12(8), 1339–1341.

Henk, J. M., Bishop, K., and Shepherd, S. F. (2003). Treatment of head and neck cancer with CHART and nimorazole: phase II study. *Radiother Oncol*, 66(1), 65–70.

Henke, M., Alfonsi, M., Foa, P. et al. (2011). Palifermin decreases severe oral mucositis of patients undergoing postoperative radiochemotherapy for head and neck cancer: a randomized, placebo-controlled trial. *J Clin Oncol*, 29(20), 2815–2820.

Heron, D. E., Rwigema, J. C., Gibson, M. K. et al. (2011). Concurrent cetuximab with stereotactic body radiotherapy for recurrent squamous cell carcinoma of the head and neck: a single institution matched case-control study. *Am J Clin Oncol*, 34, 165–172.

Herrmann, R., Fayad, W., Schwarz, S. et al. (2008). Screening for compounds that induce apoptosis of cancer cells grown as multicellular spheroids. *J Biomol Screen*, 13, 1–8.

Hicklin, D. J. and Ellis, L. M. (2005). Role of the vascular endothelial growth factor pathway in tumour growth and angiogenesis. *J Clin Oncol*, 23, 1011–1027.

Hicks, K. O., Siim, B. G., Jaiswal, J. K. et al. (2010). Pharmacokinetic/pharmacodynamic modeling identifies SN30000 and SN29751 as tirapazamine analogues with improved tissue penetration and hypoxic cell killing in tumors. *Clin Cancer Res*, 16, 4946–4957.

Hicks, K. O., Siim, B. G., Pruijn, F. B. et al. (2004). Oxygen dependence of the metabolic activation and cytotoxicity of tirapazamine: implications for extravascular transport and activity in tumors. *Radiat Res*, 161(6), 656–666.

Hill, R. P., Bristow, R. G., Fyles, A. et al. (2015). Hypoxia and predicting radiation response. *Semin Radiat Oncol*, 25(4), 260–272.

Höckel, M. and Vaupel, P. (2001). Biological consequences of tumor hypoxia. *Semin Oncol*, 28, 36–41.

Hodgkiss, R. J., Webster, L., and Wilson, G. D. (1995). Development of bioreductive markers for tumour hypoxia. *Acta Oncologica*, 34, 351–355.

Hoeben, B. A., Troost, E. G., Span, P. N. et al. (2013). 18F-FLT PET during radiotherapy or chemoradiotherapy in head and neck squamous cell carcinoma is an early predictor of outcome. *J Nucl Med*, 54(4), 532–540.

Hoffmann, A. L. and Nahum, A. E. (2013). Fractionation in normal tissues: the (α/β)eff concept can account for dose heterogeneity and volume effects. *Phys Med Biol*, 58(19), 6897–6914.

Hoffmann, T. K. (2012). Systemic therapy strategies for head-neck carcinomas: current status. *GMS Curr Top Otorhinolaryngol Head Neck Surg*, 11, Doc03.

Hoppe, B. S., Stegman, L. D., Zelefsky, M. J. et al. (2007). Treatment of nasal cavity and paranasal sinus cancer with modern radiotherapy techniques in the postoperative setting-the MSKCC experience. *Int J Radiat Oncol Biol Phys*, 67, 691–702.

Horiot, J. C., Bontemps, P., van den Bogaert, W. et al. (1997). Accelerated fractionation (AF) compared to conventional fractionation (CF) improves loco-regional control in the radiotherapy of advanced head and neck cancers: results of the EORTC 22851 randomized trial. *Radiother Oncol*, 44, 111–121.

Horiot, J. C., Bontemps, P., van den Bogaert, W. et al. (1997). Accelerated fractionation (AF) compared to conventional fractionation (CF) improves loco-regional control in the radiotherapy of advanced head and neck cancers: results of the EORTC 22851 randomized trial. *Radiother Oncol*, 44, 111–121.

Horiot, J. C., Le Fur, R., N'Guyen, T. et al. (1992). Hyperfractionation versus conventional fractionation in oropharyngeal carcinoma: final analysis of a randomized trial of the EORTC cooperative group of radiotherapy. *Radiother Oncol*, 25, 231–241.

Horsman, M. R. and Overgaard, J. (2016). The impact of hypoxia and its modification of the outcome of radiotherapy. *J Radiat Res*, 57(S1), i90–i98.

Hoshikawa, H., Mori, T., Yamamoto, Y. et al. (2015). Prognostic value comparison between (18) F-FLT PET/CT and (18)F-FDG PET/CT volume-based metabolic parameters in patients with head and neck cancer. *Clin Nucl Med*, 40(6), 464–468.

Huang, Q., Li, F., Liu, X. et al. (2011). Caspase 3-mediated stimulation of tumor cell repopulation during cancer radiotherapy. *Nat Med*, 17, 860–866.

Huang, S. D., Yuan, Y., Tang, H. et al. (2013). Tumor cells positive and negative for the common cancer stem cell markers are capable of initiating tumor growth and generating both progenies. *PLoS One*, 8, e54579.

Huang, S. M. and Harari, P. M. (2000). Modulation of radiation response after epidermal growth factor receptor blockade in squamous cell carcinomas: inhibition of damage repair, cell cycle kinetics, and tumour angiogenesis. *Clin Cancer Res*, 6(6), 2166–2174.

Hui, E. P., Chan, A. T., Pezzella, F. et al. (2002). Coexpression of hypoxia-inducible factors 1alpha and 2alpha, carbonic anhydrase IX, and vascular endothelial growth factor in nasopharyngeal carcinoma and relationship to survival. *Clin Cancer Res*, 8, 2595–2604.

Hunter, F. W., Young, R. J., Shalev, Z. et al. (2015). Identification of P450 oxidoreductase as a major determinant of sensitivity to hypoxia-activated prodrugs. *Cancer Res*, 75(19), 4211–4223.

IARC Working Group, Lyon, 23-30 October 1986: Tobacco smoking. *IARC Monogr Eval Carcinog Risk Chem Hum*, 1986;38:1–421.

IARC Working Group, Lyon 13-20 October 1987: Alcohol drinking. *IARC Monogr Eval Carcinog Risk Chem Hum*, 1988;44:1–378.

ICRU. Prescribing, recording, and reporting electron beam therapy. Bethesda: International Commission on Radiation Units and Measurements; 2004. ICRU Report 71.

ICRU. Prescribing, recording, and reporting photon beam therapy (Supplement to ICRU Report 50). Bethesda: International Commission on Radiation Units and Measurements; 1999. ICRU Report 62.

ICRU. Prescribing, recording, and reporting photon beam therapy. Bethesda: International Commission on Radiation Units and Measurements; 1993. ICRU Report 50.

ICRU. Prescribing, recording, and reporting photon-beam intensity-modulated radiation therapy (IMRT). Bethesda: International Commission on Radiation Units and Measurements; 2010. ICRU Report 83.

ICRU. Prescribing, recording, and reporting proton-beam therapy. Bethesda: International Commission on Radiation Units and Measurements; 2007. ICRU Report 78.

Jackson, S. M., Weir, L. M., Hay, J. H. et al. (1997). A randomised trial of accelerated versus conventional radiotherapy in head and neck cancer. *Radiother Oncol*, 43, 39–46.

Jain, R. (2014). Antiangiogenesis strategies revisited: from starving tumors to alleviating hypoxia. *Cancer Cell*, 26(5), 605–622.

Janot, F., de Raucourt, D., Benhamou, E. et al. (2008). Randomized trial of postoperative reirradiation combined with chemotherapy after salvage surgery compared with salvage surgery alone in head and neck carcinoma. *J Clin Oncol*, 26, 5518–5523.

Janssens, G. O., Rademakers, S. E., Terhaard, C. H. et al. (2012). Accelerated radiotherapy with carbogen and nicotinamide for laryngeal cancer: results of a phase III randomized trial. *J Clin Oncol*, 30(15), 1777–1783.

Janssens, G. O., Rademakers, S. E., Terhaard, C. H. et al. (2014). Improved recurrence-free survival with ARCON for anemic patients with laryngeal cancer. *Clin Cancer Res*, 20(5), 1345–1354.

Jemal, A., Bray, F., Center, M. M. et al. (2011). Global cancer statistics. *CA Cancer J Clin*, 61(2), 69–90.

Jeremic, B. and Milicic, B. (2008). Influence of low-dose daily cisplatin on the distant metastasis-free survival of patients with locally advanced nonmetastatic head and neck cancer treated with radiation therapy. *Radiother Oncol*, 87(2), 201–203.

Jeremic, B., Shibamoto, Y., Stanisavljevic, B. et al. (1997). Radiation therapy alone or with concurrent low-dose daily either cisplatin or carboplatin in locally advanced unresectable squamous cell carcinoma of the head and neck: a prospective randomized trial. *Radiother Oncol*, 43(1), 29–37.

Jermann, M. (2015). Particle Therapy Statistics in 2014. *Int J Particle Therapy*, 2(1), 50–54.

Joiner, M. C. and Van der Kogel, A. J. (Eds.). Basic Clinical Radiobiology, 4th edition. Boca Raton, Florida: CRC Press; 2009.

Jones, B. (2016). Why RBE must be a variable and not a constant in proton therapy. *Br J Radiol*, 1063(89), 20160116.

Jones, B., Carabe-Fernandez, A., and Dale, R. G. (2006). Calculation of high-LET radiotherapy dose required for compensation of overall treatment time extensions. *Br J Radiol*, 79(939), 254–257.

Jones, B. and Dale, R. G. (2000). Estimation of optimum dose per fraction for high LET radiations: implications for proton radiotherapy. *Int J Radiat Oncol Biol Phys*, 48(5), 1549–1557.

Jones, K. L. and Budzar, A. U. (2009). Evolving novel anti-HER 2 strategies. *Lancet Oncol*, 10, 1179–1187.

Joshua, B., Kaplan, M. J., Doweck, I. et al. (2012). Frequency of cells expressing CD44, a head and neck cancer stem cell marker: correlation with tumor aggressiveness. *Head Neck*, 34(1), 42–49.

Jun, H., Chang, M., Ko, Y. et al. (2009). Clinical significance of type 1 insulin-like growth factor receptor and insulin-like growth factor binding protein-3 expression in squamous cell carcinoma of the head and neck. *J Clin Oncol*, 27, 15s abstr 6036.

Kaanders, J. H., Pop, L., Marres, H. et al. (2002b). ARCON: experience in 215 patients with advanced head-and-neck cancer. *Int J Radiat Oncol Biol Phys*, 52(3), 769–778.

Kaanders, J. H., Pop, L. A., Marres, H. A. et al. (1995). Radiotherapy with carbogen breathing and nicotinamide in head and neck cancer: feasibility and toxicity. *Radiother Oncol*, 37, 190–198.

Kaanders, J. H., van Daal, W. A., Hoogenraad, W. J. et al. (1992). Accelerated fractionation radiotherapy for laryngeal cancer, acute, and late toxicity. *Int J Radiat Oncol Biol Phys*, 24, 497–503.

Kaanders, J. H., Wijffels, K. I., Marres, H. A. et al. (2002a). Pimonidazole binding and tumor vascularity predict for treatment outcome in head and neck cancer. *Cancer Res*, 62, 7066–7074.

Källman, P., Ågren, A., and Brahme A. (1992). Tumour and normal tissue responses to fractionated non-uniform dose delivery. *Int J Radiat Biol*, 62(2), 249–262.

Kalyankrishna, S. and Grandis, J. R. (2006). Epidermal growth factor receptor biology in head and neck cancer. *J Clin Oncol*, 17, 2666–2672.

Kao, J., Garofalo, M. C., Milano, M. T. et al. (2003). Reirradiation of recurrent and second primary head and neck malignancies: a comprehensive review. *Cancer Treat Rev*, 29, 21–30.

Karran, P. (2000). DNA double strand break repair in mammalian cells. *Curr Opin Genet Dev*, 10, 144–150.

Kase, Y., Kanai, T., Matsufuji, N. et al. (2008). Biophysical calculation of cell survival probabilities using amorphous track structure models for heavy-ion irradiation. *Phys Med Biol*, 53(1), 37–59.

Katz, T. S., Mendenhall, W. M., Morris, C. G. et al. (2002). Malignant tumors of the nasal cavity and paranasal sinuses. *Head Neck*, 24, 821–829.

Kawashiri, S., Noguchi, N., Tanaka, A. et al. (2009). Inhibitory effect of neoadjuvant chemotherapy on metastasis of oral squamous cell carcinoma in a mouse model. *Oral Oncol*, 45(9), 794–797.

Kellerer, A. M. and Rossi, H. H. (1972). The theory of dual radiation action. *Curr Top Radiat Res Q*, 8, 85–158.

Kelly, J. R., Husain, Z. A., and Burtness, B. (2016). Treatment de-intensification strategies for head and neck cancer. *Eur J Cancer*, 68, 125–133.

Kikuchi, M., Yamane, T., Shinohara, S. et al. (2011). 18F-fluoromisonidazole positron emission tomography before treatment is a predictor of radiotherapy outcome and survival prognosis in patients with head and neck squamous cell carcinoma. *Ann Nucl Med*, 25, 625–633.

Kim, K. J., Li, B., Winer, J. et al. (1993). Inhibition of vascular endothelial growth factor-induced angiogenesis suppresses tumour growth in vivo. *Nature*, 362, 841–844.

Kim, S., Loevner, L., Quon, H. et al. (2009). Diffusion-weighted magnetic resonance imaging for predicting and detecting early response to chemoradiation therapy of squamous cell carcinomas of the head and neck. *Clin Cancer Res*, 15(3), 986–994.

Kim, S. H., Koo, B. S., Kang, S. et al. (2007). HPV integration begins in the tonsillar crypt and leads to the alteration of p16, EGFR an c-myc during tumor formation. *Int J Cancer*, 120(7), 1418–1425.

King, A. and Thoeny, H. (2016). Functional MRI for the prediction of treatment response in head and neck squamous cell carcinoma: potential and limitations. *Cancer Imag*, 16, 23.

Kirk, J., Gtay, W. M., and Watson, E. R. (1971). Cumulative radiation effect: Part I. Fractionated treatment regimes. *Clin Radiol*, 22, 145–155.

Knegjens, J. L., Hauptmann, M., Pameijer, F. A. et al. (2011). Tumor volume as prognostic factor in chemoradiation for advanced head and neck cancer. *Head Neck*, 33(3), 375–382.

Knoblich, J. A. (2008). Mechanisms of asymmetric stem cell division. *Cell*, 132(4), 583–597.

Kong, C. S., Narasimhan, B., Cao, H. et al. (2009). The relationship between human papillomavirus status and other molecular prognostic markers in head and neck squamous cell carcinomas. *Int J Radiat Oncol Biol Phys*, 74, 553–561.

Koritzinsky, M., Magagnin, M. G., van den, B. T. et al. (2006). Gene expression during acute and prolonged hypoxia is regulated by distinct mechanisms of translational control. *EMBO J*, 25, 1114–1125.

Koritzinsky, M. and Wouters, B. G. (2007). Hypoxia and regulation of messenger RNA translation. *Methods Enzymol*, 435, 247–273.

Koritzinsky, M., Seigneuric, R., Magagnin, M. G. et al. (2005). The hypoxic proteome is influenced by gene-specific changes in mRNA translation. *Radiother Oncol*, 76, 177–186.

Kotwall, C., Sako, K., Razack, M. S. et al. (1997). Metastatic patterns in squamous cell cancer of the head and neck. *Am J Surg*, 154, 439–442.

Koukourakis, M. I. (2003). Amifostine: is there evidence of tumour protection? *Semin Oncol*, 30, 18–30.

Koukourakis, M. I., Giatromanolaki, A., Danielidis, V. et al. (2008). Hypoxia inducible factor (HIf1alpha and HIF2alpha) and carbonic anhydrase 9 (CA9) expression and response of head-neck cancer to hypofractionated and accelerated radiotherapy. *Int J Radiat Biol*, 84(1), 47–52.

Koukourakis, M. I., Giatromanolaki, A., Sivridis, E. et al. (2002). Hypoxia-inducible factor (HIF1A and HIF2A), angiogenesis, and chemoradiotherapy outcome of squamous cell head-and-neck cancer. *Int J Radiat Oncol Biol Phys*, 53, 1192–1202.

Koukourakis, M. I., Giatromanolaki, A., Tsakmaki, V. et al. (2012). Cancer stem cell phenotype relates to radio-chemotherapy outcome in locally advanced squamous cell head-neck cancer. *Br J Cancer*, 106(5), 846–853.

Kouvaris, J. R., Kouloulias, V. E., and Vlahos, L. J. (2007). Amifostine: the first selective-target and broad-spectrum radioprotector. *The Oncologist*, 12(6), 738–747.

Kovács, G., Martinez-Monge, R., Budrukkar, A. et al. (2017). Head & Neck Working Group GEC-ESTRO ACROP recommendations for head & neck brachytherapy in squamous cell carcinomas: 1st update—improvement by cross sectional imaging based treatment planning and stepping source technology. *Radiother Oncol*, 122(2), 248–254.

Krause, M., Yaromina, A., Eicheler, W. et al. (2011). Cancer stem cells: targets and potential biomarkers for radiotherapy. *Clin Cancer Res*, 17(23), 7224–7229.

Krishnamurthy, S., Dong, Z., Vodopyanov, D. et al. (2010). Endothelial cell-initiated signaling promotes the survival and self-renewal of cancer stem cells. *Cancer Research*, 70, 9969–9978.

Krohn, K. A., Link, J. M., and Mason, R. P. (2008). Molecular imaging of hypoxia. *J Nucl Med*, 49(Suppl 2), 129S–148S.

Kubicek, G., Axelrod, R., Machtay, M. et al. (2012). Phase I trial using proteasome inhibitor bortezomib and concurrent chemoradiotherapy for head-and-neck malignancies. *Int J Rad Oncol Biol Phys*, 83(4), 1192–1197.

Kumareswaran, R., Ludkovski, O., Meng, A. et al. (2012). Chronic hypoxia compromises repair of DNA double-strand breaks to drive genetic instability. *J Cell Sci*, 125, 189–199.

Kurth, I., Hein, L., Mäbert, K. et al. (2015). Cancer stem cell related markers of radioresistance in head and neck squamous cell carcinoma. *Oncotarget*, 6(33), 34494–34509.

Kusumbe, A. P. and Bapat, S. A. (2009). Cancer stem cells and aneuploid populations within developing tumors are the major determinants of tumor dormancy. *Cancer Res*, 69(24), 9245–9253.

Kutler, D. I., Wreesmann, J. B., Goberdhan, A. et al. (2003). Human papillomavirus DNA and p53 polymorphisms in squamous cell carcinomas from Fanconi anemia patients. *J Natl Cancer Inst*, 95(22), 1718–1721.

Kwon, O. J., Park, J. J., Ko, G. H. et al. (2015). HIF-1α and CA-IX as predictors of locoregional control for determining the optimal treatment modality for early-stage laryngeal carcinoma. *Head Neck*, 37(4), 505–510.

Lacas, B., Bourhis, J., Overgaard, J. et al. (2017). Role of radiotherapy fractionation in head and neck cancers (MARCH): an updated meta-analysis. *Lancet Oncol*, 18, 1221–1237.

Langendijk, J. A., Kasperts, N., Leemaans, C. R. et al. (2006). A phase II study of primary reirradiation in squamous cell carcinoma of head and neck. *Radiother Oncol*, 78, 306–312.

Langlois, D., Hoffstetter, S., Malissard, L. et al. (1988). Salvage irradiation of oropharynx and mobile tongue about 192 iridium brachytherapy in Centre Alexis Vautrin. *Int J Radiat Oncol Biol Phys*, 14, 849–853.

Laramore, G. E. (2009). Role of particle radiotherapy in the management of head and neck cancer. *Curr Opin Oncol*, 21(3), 224–231.

Laramore, G. E., Krall, J. M., Griffin, T. W. et al. (1993). Neutron versus photon irradiation for unresectable salivary gland tumours; final report of a RTOG randomized trial. *Int J Radiat Oncol Biol Phys*, 27, 2235–2240.

Lartigau, E. F., Tresch, E., Thariat, J. et al. (2013). Multi institutional phase II study of concomitant stereotactic reirradiation and cetuximab for recurrent head and neck cancer. *Radiother Oncol*, 109, 281–285.

Lassen, P., Eriksen, J. G., Hamilton-Dutoit, S. et al. (2010). HPV-associated p16-expression and response to hypoxic modification of radiotherapy in head and neck cancer. *Radiother Oncol*, 94(1), 30–35.

Lassen, P., Primdahl, H., Johansen, J. et al. (2014). Impact of HPV-associated p16-expression on radiotherapy outcome in advanced oropharynx and non-oropharynx cancer. *Radiother Oncol*, 113(3), 310–316.

Lawrence, T. S., Blackstock, A. W., and McGinn, C. (2003). The mechanism of action of radiosensitization of conventional chemotherapeutic agents. *Sem Radiat Oncol*, 13(1), 13–21.

Le, Q. T., Fisher, R., Oliner, K. S. et al. (2012). Prognostic and predictive significance of plasma HGF and IL-8 in a phase III trial of chemoradiation with or without tirapazamine in locoregionally advanced head and neck cancer. *Clin Cancer Res*, 18(6), 1798–1807.

Le, Q. T., Kim, H. E., Schneider, C. J. et al. (2011). Palifermin reduces severe mucositis in definitive chemoradiotherapy of locally advanced head and neck cancer: a randomized, placebo-controlled study. *J Clin Oncol*, 29(20), 2808–2814.

Lea, D. E. Actions of Radiations on Living Cells. Cambridge: The University Press; 1946.

Lea, D. E. and Catcheside D. G. (1942). The mechanism of the induction by radiation of chromosome aberrations in Tradescantia. *J Genet*, 44, 216–245.

Lee, D. J., Cosmatos, D., Marcial, V. A. et al. (1995). Results of an RTOG phase III trial (RTOG 85-27) comparing radiotherapy plus etanidazole with radiotherapy alone for locally advanced head and neck carcinomas. *Int J Radiat Oncol Biol Phys*, 32, 567–576.

Lee, N., Chan, K., Bekelman, J. E. et al. (2007). Salvage re-irradiation for recurrent head and neck cancer. *Int J Radiat Oncol Biol Phys*, 68, 731–740.

Lee, N., Nehmeh, S., Schöder, H. et al. (2009). Prospective trial incorporating pre-/mid-treatment [18F]-misonidazole positron emission tomography for head-and-neck cancer patients undergoing concurrent chemoradiotherapy. *Int J Radiat Oncol Biol Phys*, 75, 101–108.

Lee, S. H., Koo, B. S., Kim, J. M. et al. (2014). Wnt/β-catenin signalling maintains self-renewal and tumourigenicity of head and neck squamous cell carcinoma stem-like cells by activating Oct4. *J Pathol*, 234(1), 99–107.

Lee, S. T. and Scott, A. M. (2007). Hypoxia positron emission tomography imaging with 18f-fluoromisonidazole. *Semin Nucl Med*, 37, 451–461.

Leeman, J. E., Romesser, P. B., Zhou, Y. et al. (2017). Proton therapy for head and neck cancer: expanding the therapeutic window. *Lancet Oncol*, 18(5), e254–e265.

Lesher, S. and Baumann, J. (1969). Cell kinetic studies of the intestinal epithelium: maintenance of the intestinal epithelium in normal and irradiated animals. *Natl Cancer Inst Monogr*, 30, 185–198.

Leung, D. W., Cachianes, G., Kuang, W. J. et al. (1989). Vascular endothelial growth factor is a secreted angiogenic mitogen. *Science*, 246(4935), 1306–1309.

Lewis, J. S. Jr, Ukpo, O. C., Ma, X. J. et al. (2012). Transcriptionally-active high-risk human papillomavirus is rare in oral cavity and laryngeal/hypopharyngeal squamous cell carcinomas—a tissue microarray study utilizing E6/E7 mRNA in situ hybridization. *Histopathology*, 60, 982–991.

Lewis, J. S., McCarthy, D. W., McCarthy, T. J. et al. (1999). Evaluation of 64Cu-ATSM in vitro and in vivo in a hypoxic tumor model. *J Nucl Med*, 40, 177–183.

Lieber, M. R., Ma, Y., Pannicke, U. et al. (2003). Mechanism and regulation of human non-homologous DNA end-joining. *Nat Rev Mol Cell Biol*, 4, 712–720.

Liebig, H., Günther, G., Kolb, M. et al. (2017). Reduced proliferation and colony formation of head and neck squamous cell carcinoma (HNSCC) after dual targeting of EGFR and hedgehog pathways. *Cancer Chemother Pharmacol*, 79(2), 411–420.

Lievens, Y., Haustermans, K., Van den Weyngaert, D. et al. (1998). Does sucralfate reduce the acute side-effects in head and neck cancer treated with radiotherapy? A double-blind randomized trial. *Radiother Oncol*, 47, 149–153.

Lim, Y. C., Kang, H. J., Kim, Y. S. et al. (2012). All-trans-retinoic acid inhibits growth of head and neck cancer stem cells by suppression of Wnt/β-catenin pathway. *Eur J Cancer*, 48, 3310–3318.

Lindegaard, J. (2003). Has the time come for routine use of amifostine in clinical radiotherapy practice? *Acta Oncol*, 42, 2–3.

Lindquist, D., Ahrlund-Richter, A., Tarján, M. et al. (2012). Intense CD44 expression is a negative prognostic factor in tonsillar and base of tongue cancer. *Anticancer Res*, 32, 153–161.

Ling, C. C., Humm, J., Larson, S. et al. (2000). Towards multidimensional radiotherapy (MD-CRT): biological imaging and biological conformality. *International Journal of Radiation Oncology Biology Physics*, 47(3), 551–560.

Linge, A., Lohaus, F., Löck, S. et al. (2016). HPV status, cancer stem cell marker expression, hypoxia gene signatures and tumour volume identify good prognosis subgroups in patients with HNSCC after primary radiochemotherapy: a multicentre retrospective study of the German Cancer Consortium Radiation Oncology Group (DKTK-ROG). *Radiother Oncol*, 121(3), 364–373.

Lingen, M. W., Xiao, W., Schmitt, A. et al. (2013). Low etiologic fraction for high-risk human papillomavirus in oral cavity squamous cell carcinomas. *Oral Oncol*, 49, 1–8.

Liu, J., Pan, S., Hsieh, M. H. et al. (2013). Targeting Wnt-driven cancer through inhibition of Porcupine by LGK974. *Proc Natl Acad Sci USA*, 110(50), 20224–20229.

Liu, Z., Liu, J., Li, L. et al. (2015). Inhibition of autophagy potentiated the antitumor effect of nedaplatin in cisplatin-resistant nasopharyngeal carcinoma cells. *PLoS One*, 10(8), e0135236.

Ljungkvist, A., Bussink, J., Kaanders, J. et al. (2005). Hypoxic cell turnover in different solid tumor lines. *Int J Rad Oncol Biol Phys*, 62, 1157–1168.

Ljungkvist, A., Bussink, J., Kaanders, J. et al. (2006). Dynamics of hypoxia, proliferation and apoptosis after irradiation in a murine tumor model. *Rad Res*, 165, 326–336.

Ljungkvist, A., Bussink, J., Rijken, P. F. J. W. et al. (2002). Vascular architecture, hypoxia, and proliferation in first-generation xenografts of human head-and-neck squamous cell carcinomas. *Int J Radiat Oncol Biol Phys*, 54, 215–228.

Lo, K. W., Chung, G. T., To, and K. F. (2012). Deciphering the molecular genetic basis of NPC through molecular, cytogenetic, and epigenetic approaches. *Semin Cancer Biol*, 22(2), 79–86.

Lo Nigro, C., Denaro, N., Merlotti, A. et al. (2017). Head and neck cancer: improving outcomes with a multidisciplinary approach. *Cancer Manag Res*, 9, 363–371.

Loges, S., Schmidt, T., and Carmeliet, P. (2010). Mechanisms of resistance to anti-angiogenic therapy and development of third-generation anti-angiogenic drug candidates. *Genes Cancer*, 1(1), 12–25.

Lombaert, I. M., Brunsting, J. F., Wierenga, P. K. et al. (2008). Keratinocyte growth factor prevents radiation damage to salivary glands by expansion of the stem/progenitor pool. *Stem Cells*, 26(10), 2595–2601.

Lowy, D. R. and Gillison, M. L. (2003). A new link between Fanconi anemia and human papillomavirus-associated malignancies. *J Natl Cancer Inst*, 95(22), 1648–1650.

Lun, M., Zhang, P. L., Pellitteri, P. K. et al. (2005). Nuclear factor-kappa-B pathway as a therapeutic target in head and neck squamous cell carcinoma: pharmaceutical and molecular validation in human cell lines using Velcade and siRNA/NFkappaB. *Ann Clin Lab Sci*, 35, 251–258.

Luoto, K. R., Kumareswaran, R., and Bristow, R. G. (2013). Tumor hypoxia as a driving force in genetic instability. *Genome Integr*, 4:5.

Lydiatt, W. M., Shah, J. P., and Hoffman, H. T. (2001). American Joint Committee on Cancer. AJCC stage groupings for head and neck cancer: should we look at alternatives? A report of the Head and Neck Sites Task Force. *Head Neck*, 23, 607–612.

Mach, R. H., Dehdashti, F., and Wheeler, K. T. (2009). PET radiotracers for imaging the proliferative status of solid tumors. *PET Clin*, 4(1), 1–15.

Macha, M. A., Rachagani, S., Qazi, A. K. et al. (2017). Afatinib radiosensitizes head and neck squamous cell carcinoma cells by targeting cancer stem cells. *Oncotarget*, 8(13), 20961–20973.

Machiels, J. P., Haddad, R. I., Fayette, J. et al. (2015). Afatinib versus methotrexate as second-line treatment in patients with recurrent or metastatic squamous-cell carcinoma of the head and neck progressing on or after platinum-based therapy (LUX-Head & Neck 1): an open-label, randomised phase 3 trial. *Lancet Oncol*, 16(5), 583–594.

Machiels, J. P., Henry, S., Zanetta, S. et al. (2010). Phase II study of sunitinib in recurrent or metastatic squamous cell carcinoma of the head and neck: GORTEC 2006-01. *J Clin Oncol*, 28(1), 21–28.

Machiels, J. P., Subramanian, S., Ruzsa, A. et al. (2011). Zalutumumab plus best supportive care versus best supportive care alone in patients with recurrent or metastatic squamous-cell carcinoma of the head and neck after failure of platinum-based chemotherapy: an open-label, randomised phase 3 trial. *Lancet Oncol*, 12(4), 333–343.

Maciejewski, B., Skladowski, K., Pilecki, B. et al. (1996). Randomized clinical trial on accelerated 7 days per week fractionation in radiotherapy for head and neck cancer: preliminary report on acute toxicity. *Radiother Oncol*, 40, 137–145.

Maciejewski, B., Zajusz, A., Pilecki, B. et al. (1991). Acute mucositis in the stimulated oral mucosa of patients during radiotherapy for head and neck cancer. *Radiother Oncol*, 22, 7–11.

Maier, H., Dietz, A., Gewelke, U. et al. (1992). Tobacco and alcohol and the risk of head and neck cancer. *Clin Investig*, 70(3–4), 320–327.

Maisin, J. R. (1989). Chemical protection against ionizing radiation. *Advances in Space Research*, 9, 205–212.

Majem, M., Mesia, R., Mañós, M. et al. (2006). Does induction chemotherapy still have a role in larynx preservation strategies? The experience of Institut Catala d'Oncologia in stage III larynx carcinoma. *Laryngoscope*, 116, 1651–1656.

Malm, I. J., Fan, C. J., Yin, L. X. et al. (2017). Evaluation of proposed staging systems for human papillomavirus-related oropharyngeal squamous cell carcinoma. *Cancer*, 123(10), 1768–1777.

Mandal, M., Myers, J. N., Lippman, S. M. et al. (2008). Epithelial to mesenchymal transition in head and neck squamous carcinoma: association of Src activation with E-cadherin down-regulation, vimentin expression, and aggressive tumour features. *Cancer*, 112, 2088–2100.

Marcu, L. and Bezak, E. (2012). The influence of stem cell cycle time on accelerated repopulation during radiotherapy in head and neck cancer. *Cell Prolif*, 45(5), 404–412.

Marcu, L., Bezak, E., and Filip, S. (2012). The role of PET imaging in overcoming radiobiological challenges in the treatment of advanced head and neck cancer. *Cancer Treat Rev*, 8(3), 185–193.

Marcu, L., Harriss-Phillips, W., and Filip, S. (2014). Hypoxia in head and neck cancer in theory and practice: a PET-based imaging approach. *Comput Mathem Meth Med*, 2014, 624642.

Marcu, L., van Doorn, T., and Olver, I. (2003). Cisplatin and radiotherapy in the treatment of locally advanced head and neck cancer. *Acta Oncol*, 42(4), 315–325.

Marcu, L., van Doorn, T., and Olver, I. (2004). Modelling of post irradiation accelerated repopulation in squamous cell carcinomas. *Phys Med Biol*, 49, 3676–3779.

Marcu, L. and Yeoh, E. Tumour repopulation during treatment for head and neck cancer: clinical evidence, mechanisms and minimizing strategies. In: Agulnik, M. (Ed.). Head and Neck Cancer. InTech Publishing; 2012.

Marcu, L. G. (2009). The role of amifostine in the treatment of head and neck cancer with cisplatin-radiotherapy. *Eur J Cancer Care*, 18(2), 116–123.

Marcu, L. G. (2013). Improving therapeutic ratio in head and neck cancer with adjuvant and cisplatin-based treatments. *Biomed Res Int*, 2013, 817279.

Marcu, L. G. (2014). Tumour repopulation and the role of abortive division in squamous cell carcinomas during chemotherapy. *Cell Prolif*, 47(4), 318–325.

Marcu, L. G. (2016). Future treatment directions for HPV-associated head and neck cancer based on radiobiological rationale and current clinical evidence. *Critical Rev Oncol/Hematol*, 103, 27–36.

Marcu, L. G. and Marcu, D. (2016). In silico modelling of a cancer stem cell-targeting agent and its effects on tumour control during radiotherapy. *Sci Reports*, 6, 32332.

Marks, L. B., Yorke, E. D., Jackson, A. et al. (2010). Use of normal tissue complication probability models in the clinic. *Int J Radiat Oncol Biol Phys*, 76, S10–S19.

Martin, G. S. (2001). The hunting of the Src. *Nat Rev Mol Cell Biol*, 2, 467–475.

Martins, R. G., Parvathaneni, U., Bauman, J. E. et al. (2013). Cisplatin and radiotherapy with or without erlotinib in locally advanced squamous cell carcinoma of the head and neck: a randomized phase II trial. *J Clin Oncol,* 31(11), 1415–1421.

Mayer, R., Hamilton-Farrell, M. R., van der Kleij, A. J. et al. (2005). Hyperbaric oxygen and radiotherapy. *Strahlenther Onkol,* 181(2), 113–123.

Mazeron, J. J., Ardiet, J. M., Haie-Méder, C. et al. (2009). GEC-ESTRO recommendations for brachytherapy for head and neck squamous cell carcinomas. *Radiother Oncol,* 91(2), 150–156.

McGurk, M. and Goodger, N. M. (2000). Head and neck cancer and its treatment: historical review. *Br J Oral Maxillofac Surg,* 38(3), 209–220.

McQuestion, M. (2011). Evidence-based skin care management in radiation therapy: clinical update. *Semin Oncol Nurs,* 27, e1–e17.

McWhinney, S. R., Goldberg, R. M., and McLeod, H. L. (2009). Platinum neurotoxicity pharmacogenetics. *Mol Cancer Ther,* 8(1), 10–16.

Menda, Y., Boles Ponto, L. L., Dornfeld, K. J. et al. (2009). Kinetic analysis of 3'-deoxy-3'-(18) F-fluorothymidine ((18)F-FLT) in head and neck cancer patients before and early after initiation of chemoradiation therapy. *J Nucl Med,* 50(7), 1028–1035.

Mendelsohn, J. (1997). Epidermal growth factor receptor inhibition by a monoclonal antibody as anticancer therapy. *Clin Cancer Res,* 3, 2703–2707.

Mendelsohn, J. and Baselga, J. (2000). The EGF receptor family as targets for cancer therapy. *Oncogene,* 19, 6550–6565.

Mendenhall, W. M., Mancuso, A. A., Strojan, P. et al. (2014). Impact of primary tumor volume on local control after definitive radiotherapy for head and neck cancer. *Head Neck,* 36(9), 1363–1367.

Meng, A. X., Jalali, F., Cuddihy, A. et al. (2005). Hypoxia down-regulates DNA double strand break repair gene expression in prostate cancer cells. *Radiother Oncol,* 76, 168–176.

Meredith, R., Salter, M., Kim, R. et al. (1997). Sucralfate for radiation mucositis: results of a double-blind randomized trial. *Int J Radiat Oncol Biol Phys,* 37, 275–279.

Mesia, R., Henke, M., Fortin, A. et al. (2015). Chemoradiotherapy with or without panitumumab in patients with unresected, locally advanced squamous-cell carcinoma of the head and neck (CONCERT-1): a randomised, controlled, open-label phase 2 trial. *Lancet Oncol,* 16(2), 208–220.

Miah, A. B., Schick, U., Bhide, S. A. et al. (2015). A phase II trial of induction chemotherapy and chemo-IMRT for head and neck squamous cell cancers at risk of bilateral nodal spread: the application of a bilateral superficial lobe parotid-sparing IMRT technique and treatment outcomes. *Br J Cancer,* 112(1), 32–38.

Michael, B. D., Adams, C. E., Hewitt, H. B. et al. (1973). A posteffect of oxygen in irradiated bacteria: a submillisecond fast mixing study. *Radiat Res,* 54, 239–251.

Milas, L., Mason, K., Liao, Z. et al. (2003). Chemoradiotherapy: emerging treatment improvement strategies. *Head Neck,* 25, 152–167.

Min, K. W. and Houck, J. R. (1998). Protocol for the examination of specimens from patients with carcinomas of the upper aerodigestive tract: carcinomas of the oral cavity including lip and tongue, nasal and paranasal sinuses, pharynx, larynx, salivary glands, hypopharynx, oropharynx, and nasopharynx. Cancer Committee, College of American Pathologists. *Arch Pathol Lab Med,* 122, 222–230.

Momm, F., Bartelt, S., Haigis, K. et al. (2005). Spectrophotometric skin measurements correlate with EORTC/RTOG-common toxicity criteria. *Strahlenther Onkol,* 181, 392–395.

Montejo, M. E., Shrieve, D. C., Bentz, B. G. et al. (2011). IMRT with simultaneous integrated boost and concurrent chemotherapy for locoregionally advanced squamous cell carcinoma of the head and neck. *Int J Radiat Oncol Biol Phys,* 81, e845–e852.

Moore, N. and Lyle, S. (2011). Quiescent, slow-cycling stem cell populations in cancer: a review of the evidence and discussion of significance. *J Oncol,* Pii, 396076.

Mork, J., Lie, A. K., Glattre, E. et al. (2001). Human papillomavirus infection as a risk factor for squamous cell carcinoma of the head and neck. *N Engl J Med,* 344(15), 1125–1131.

Morrison, S. J. and Kimble, J. (2006). Asymmetric and symmetric stem-cell division in development and cancer. *Nature,* 441(7097), 1068–1074.

Mortensen, L. S., Johansen, J., Kallehauge, J. et al. (2012). FAZA PET/CT hypoxia imaging in patients with squamous cell carcinoma of the head and neck treated with radiotherapy: results from the DAHANCA 24 trial. *Radiother Oncol,* 105(1), 14–20.

Mottram, J. C. (1936). A factor of importance in the radio sensitivity of tumours. *Br J Radiol,* 9, 606–614.

Mould, R. A Century of X-rays and Radioactivity in Medicine. Bristol: Institute of Physics Publishing; 1993.

Muller, J. M., Chevrier, L., Cochaud, S. et al. (2007). Hedgehog, Notch and Wnt developmental pathways as targets for anti-cancer drugs. *Drug Discov Today: Dis Mech,* 4(4), 285–291.

Müller, T., Braun, M., Dietrich, D. et al. (2017). PD-L1: a novel prognostic biomarker in head and neck squamous cell carcinoma. *Oncotarget,* 8(32), 52889–52900.

Munro, T. R. (1970). The relative radiosensitivity of the nucleus and cytoplasm of Chinese hamster fibroblasts. *Radiat Res,* 42, 451–470.

Myoui, A., Nishimura, R., Williams, P. et al. (2003). The molecule c-Src tyrosine kinase activity is associated with tumour colonization in bone and lung in an animal model of human breast cancer metastasis. *Cancer Res,* 63, 5028–5033.

Nagle, W. A., Moss, A. J. Jr., Roberts, H. G. Jr. (1980). Effects of 5-thio-D-glucose on cellular adenosine triphosphate levels and deoxyribonucleic acid rejoining hypoxic and aerobic Chinese hamster cells. *Radiology,* 137, 203–211.

Näsman, A., Attner, P., Hammarstedt, L. et al. (2009). Incidence of human papillomavirus (HPV) positive tonsillar carcinoma in Stockholm, Sweden: an epidemic of viral-induced carcinoma? *Int J Cancer,* 125, 362–366.

Naumov, G. N., Nilsson, M. B., Cascone, T. et al. (2009). Combined vascular endothelial growth factor receptor and epidermal growth factor receptor (EGFR) blockade inhibits tumour growth in xenograft models of EGFR inhibitor resistance. *Clin Cancer Res,* 15, 3484–3494.

Nauta, J. M., van Leengoed, H. L., Star, W. M. et al. (1996). Photodynamic therapy of oral cancer. A review of basic mechanisms and clinical applications. *Eur J Oral Sci,* 104, 69–81.

Nevins, J. R. (2001). The Rb/E2F pathway and cancer. *Hum Mol Genet,* 10(7), 699–703.

Newbold, K., Castellano, I. (2009). Charles-Edwards, E. et al. An exploratory study into the role of dynamic contrast-enhanced magnetic resonance imaging or perfusion computed tomography for detection of intratumoral hypoxia in head-and-neck cancer. *Int J Radiat Oncol Biol Phys,* 74(1), 29–37.

Nichols, A. C., Kneuertz, P. J., Descler, D. G. et al. (2011). Surgical salvage of the oropharynx after failure of organ-sparing therapy. *Head Neck,* 33, 516–524.

Nieder, C., Milas, L., and Ang, K. K. (2000). Tissue tolerance to reirradiation. *Semin Radiat Oncol,* 10, 200–209.

Nilsson, P., Thames, H. D., and Joiner, M. C. (1990). A generalized formulation of the 'incomplete-repair' model for cell survival and tissue response to fractionated low dose-rate irradiation. *Int J Radiat Biol,* 57(1), 127–142.

Nordsmark, M. and Overgaard, J. (2004). Tumor hypoxia is independent of hemoglobin and prognostic for loco-regional tumor control after primary radiotherapy in advanced head and neck cancer. *Acta Oncol,* 43(4), 396–403.

Nordsmark, M., Bentzen, S. M., Rudat, V. et al. (2005). Prognostic value of tumor oxygenation in 397 head and neck tumors after primary radiation therapy. An international multi-center study. *Radiother Oncol,* 77(1), 18–24.

Nordsmark, M., Overgaard, M., and Overgaard, J. (1996). Pretreatment oxygenation predicts radiation response in advanced squamous cell carcinoma of the head and neck. *Radiother Oncol,* 41, 31–39.

Nutting, C. M., Morden, J. P., Harrington, K. J. et al. (2011). Parotid-sparing intensity modulated versus conventional radiotherapy in head and neck cancer (PARSPORT): a phase 3 multicentre randomised controlled trial. *Lancet Oncol,* 12(2), 127–136.

Nyflot, M. J., Kruser, T. J., Traynor, A. M. et al. (2015). Phase 1 Trial of bevacizumab with concurrent chemoradiation therapy for squamous cell carcinoma of the head and neck with exploratory functional imaging of tumor hypoxia, proliferation, and perfusion. *Int J Radiat Oncol Biol Phys,* 91(5), 942–951.

Nylander, K., Dabelsteen, E., and Hall, P. A. (2000). The p53 molecule and its prognostic role in squamous cell carcinomas of the head and neck. *J Oral Pathol Med,* 29, 413–425.

Nystrom, J., Geladi, P., Lindholm-Sethson, B. et al. (2004). Objective measurements of radiotherapy-induced erythema. *Skin Res Technol,* 10, 242–250.

O' Sullivan, B., Huang, S. H., Su, J. et al. (2016). Development and validation of a staging system for HPV-related oropharyngeal cancer by the International Collaboration on Oropharyngeal cancer Network for Staging (ICON-S): a multicentre cohort study. *Lancet Oncol,* 17(4), 440–451.

O'Driscoll, M. and Jeggo, P. A. (2006). The role of double-strand break repair—insights from human genetics. *Nat Rev Genet,* 7, 45–54.

Okuno, S. H., Foote, R. L., Loprinzi, C. L. et al. (1997). A randomized trial of a nonabsorbable antibiotic lozenge given to alleviate radiation-induced mucositis. *Cancer,* 79, 2193–2199.

Orton, C. G. and Ellis, F. (1973). A simplification in the use of the NSD concept in practical radiotherapy. *Brit J Radiol,* 46, 529–537.

Overgaard, J. (1994). Clinical evaluation of nitroimidazoles as modifiers of hypoxia in solid tumors. *Oncol Res,* 6, 509–518.

Overgaard, J. (2007). Hypoxic radiosensitization: adored and ignored. *J Clin Oncol,* 25, 4066–4074.

Overgaard, J. (2011). Hypoxic modification of radiotherapy in squamous cell carcinoma of the head and neck-a systematic review and meta-analysis. *Radiother Oncol,* 100(1), 22–32.

Overgaard, J. and Horsman, M. R. (1996). Modification of hypoxia-induced radioresistance in tumors by the use of oxygen and sensitizers. *Semin Radiat Oncol,* 6, 10–21.

Overgaard, J., Hansen, H. S., Andersen, A. P. et al. (1998a). Misonidazole combined with split-course radiotherapy in the treatment of invasive carcinoma of larynx and pharynx: report from the DAHANCA 2 study. *Int J Radiat Oncol Biol Phys,* 16(4), 1065–1068.

Overgaard, J., Hansen, H. S., Overgaard, M. et al. (1998b). A randomized double-blind phase III study of nimorazole as a hypoxic radiosensitizer of primary radiotherapy in supraglottic larynx and pharynx carcinoma. Results of the Danish Head and Neck Cancer Study (DAHANCA) Protocol 5-85. *Radiother Oncol,* 46(2), 135–146.

Overgaard, J., Hansen, H. S., Specht, L. et al. (2003). Five compared with six fractions per week of conventional radiotherapy of squamous-cell carcinoma of head and neck: DAHANCA 6 and 7 randomised controlled trial. *Lancet,* 362, 933–940.

Overgaard, J., Sand Hansen, H., Lindeløv, B. et al. (1991). Nimorazole as a hypoxic radiosensitizer in the treatment of supraglottic larynx and pharynx carcinoma. First report from the Danish Head and Neck Cancer Study (DAHANCA) protocol 5-85. *Radiother Oncol,* 20(Suppl 1), 143–149.

Owen, D., Iqbal, F., and Pollock, B. E. (2014). Long term follow up of stereotactic radiosurgery for head and neck malignancies. *Head Neck,* 37(11), 1557–1562.

Pandor, A., Eggington, S., Paisley, S. et al. (2006). The clinical and cost-effectiveness of oxaliplatin and capecitabine for the adjuvant treatment of colon cancer: systematic review and economic evaluation. *Health Technol Assess,* 10, 1–204.

Parkin, D. M., Whelan, S. L., Ferlay, J. et al. Cancer Incidence in Five Continents. IARC Scientific Publication; 2002. Vol VIII.

Patt, H. M., Tyree, E. B., Straub, R. L. et al. (1949). Cysteine protection against x irradiation. *Science*, 110, 213–214.

Pavy, J. J., Denekamp, J., Letschert, J. et al. (1995). EORTC Late Effects Working Group. Late effects toxicity scoring: the SOMA scale. *Radiother Oncol*, 35, 11–15.

Paximadis, P., Yoo, G., Lin, H. S. et al. (2012). Concurrent chemoradiotherapy improves survival in patients with hypopharyngeal cancer. *Int J Radiat Oncol Biol Phys*, 82(4), 1515–1521.

Peng, P. J., Ou, X. Q., Chen, Z. B. et al. (2013). Multicenter phase II study of capecitabine combined with nedaplatin for recurrent and metastatic nasopharyngeal carcinoma patients after failure of cisplatin-based chemotherapy. *Cancer Chemother Pharmacol*, 72, 323–328.

Peters, L. J. and Withers, H. R. (1997). Applying radiobiological principles to combined modality treatment of head and neck cancer—the time factor. *Int J Rad Oncol Biol Phys*, 39, 831–836.

Peters, L. J. and Withers, H. R. (1997). Applying radiobiological principles to combined modality treatment of head and neck cancer-the time factor. *Int J Radiat Oncol Biol Phys*, 39(4), 831–836.

Pettersen, E. O. and Wang, H. (1996). Radiation-modifying effect of oxygen in synchronized cells pre-treated with acute or prolonged hypoxia. *Int J Radiat Biol*, 70, 319–326.

Pfister, D., Lee, N., Sherman, E. et al. (2009). Phase II study of bevacizumab (B) plus cisplatin (C) plus intensity-modulated radiation therapy (IMRT) for locoregionally advanced head and neck squamous cell cancer (HNSCC): preliminary results. *J Clin Oncol (Meeting Abstracts)*, 27, 6013.

Phillips, T. M., McBride, W. H., and Pajonk, F. (2006). The response of CD24(-/low)/CD44+ breast cancer-initiating cells to radiation. *J Natl Cancer Inst*, 98, 1777–1785.

Piert, M., Machulla, H. J., Picchio, M. et al. (2005). Hypoxia-specific tumor imaging with 18F-fluoroazomycin arabinoside. *J Nucl Med*, 46, 106–113.

Pignon, J. P., Bourhis, J., Domenge, C. et al. (2000). Chemotherapy added to locoregional treatment for head and neck squamous-cell carcinoma: three meta-analyses of updated individual data. MACH-NC Collaborative Group. Meta-analysis of chemotherapy on head and neck cancer. *Lancet*, 355, 949–955.

Pignon, J. P., Maitre, A. L., Maillard, E. et al. (2009). Meta-analysis of chemotherapy in head and neck cancer (MACH-NC): an update on 93 randomised trials and 17,346 patients. *Radiother Oncol*, 92(1), 4–14.

Platteaux, N., Dirix, P., Vanstraelen, V. et al. (2011). Outcome after re-irradiation of head and neck cancer patients. *Strahlenther Onkol*, 1, 23–31.

Pointreau, Y., Azzopardi, N., Ternant, D. et al. (2016). Cetuximab pharmacokinetics influences overall survival in head and neck cancer patients. *Ther Drug Monit*, 38(5), 567–572.

Pollard, J. D., Hanasono, M. M., Mikulec, A. A. et al. (2000). Head and neck cancer in cardiothoracic transplant recipients. *Laryngoscope*, 110(8), 1257–1261.

Posner, M. R., Lorch, J. H., Goloubeva, O. et al. (2011). Survival and human papillomavirus in oropharynx cancer in TAX 324: a subset analysis from an international phase III trial. *Ann Oncol*, 22(5), 1071–1077.

Poulsen, M. G., Denham, J. W., Peters, L. J. et al. (2001). A randomised trial of accelerated and conventional radiotherapy for stage III and IV squamous carcinoma of the head and neck: a Trans-Tasman Radiation Oncology Group Study. *Radiother Oncol*, 60, 113–122.

Prazmo, E. J., Kwasny, M., Lapinski, M. et al. (2016). Photodynamic therapy as a promising method used in the treatment of oral diseases. *Adv Clin Exp Med*, 25(4), 799–807.

Preciado, D. A., Matas, A., and Adams, G. L. (2002). Squamous cell carcinoma of the head and neck in solid organ transplant recipients. *Head Neck*, 24(4), 319–325.

Prince, M. E., Sivanandan, R., Kaczorowski, A. et al. (2007). Identification of a subpopulation of cells with cancer stem cell properties in head and neck squamous cell carcinoma. *Proc Natl Acad Sci USA*, 104, 973–978.

Prise, K. M., Gillies, N. E., and Michael, B. D. (1999). Further evidence for double-strand breaks originating from a paired radical precursor from studies of oxygen fixation processes. *Radiat Res,* 151, 635–641.

Psyrri, A. and Dafni, U. (2014). Combining cetuximab with chemoradiotherapy in locally advanced head and neck squamous cell carcinoma: is more better? *J Clin Oncol,* 32(27), 2929–2931.

Puck, T. T. and Marcus, P. I. (1956). Action of x-rays on mammalian cells. *J Exp Med,* 103(5), 653–666.

Radiation Therapy With Cisplatin or Cetuximab in Treating Patients With Oropharyngeal Cancer. Available from https://clinicaltrials.gov/show/NCT01302834. (Accessed on 16/01/2017.)

Rajarajan, A., Stokes, A., Bloor, B. K. et al. (2012). CD44 expression in oro-pharyngeal carcinoma tissues and cell lines. *PLoS One,* 7, e28776.

Rajendran, J. G., Schwartz, D. L., O'Sullivan, J. et al. (2006). Tumor hypoxia imaging with [F-18] fluoromisonidazole positron emission tomography in head and neck cancer. *Clin Cancer Res,* 12, 5435–5441.

Raju, U., Molkentine, D. P., Valdecanas, D. R. et al. (2015). Inhibition of EGFR or IGF-1R signaling enhances radiation response in head and neck cancer models but concurrent inhibition has no added benefit. *Cancer Medicine,* 4(1), 65–74.

Raju, U., Riesterer, O., and Wang, Z. (2012). Dasatinib, a multi-kinase inhibitor increased radiation sensitivity by interfering with nuclear localization of epidermal growth factor receptor and by blocking DNA repair pathways. *Radiother Oncol,* 105, 241–249.

Raleigh, J. A., Chou, S. C., Calkins-Adams, D. P. et al. (2000). A clinical study of hypoxia and metallothionein protein expression in squamous cell carcinomas. *Clin Cancer Res,* 6, 855–862.

Rasey, J. S., Grunbaum, Z., Magee, S. et al. (1987). Characterization of radiolabeled fluoromisonidazole as a probe for hypoxic cells. *Radiat Res,* 111, 292–304.

Rawat, S., Ahlawat, P., Kakria, A. et al. (2017). Comparison between weekly cisplatin-enhanced radiotherapy and cetuximab-enhanced radiotherapy in locally advanced head and neck cancer: first retrospective study in Asian population. *Asia Pac J Clin Oncol,* 13(3), 195–203.

Reddy, B. K., Lokesh, V., Vidyasagar, M. S. et al. (2014). Nimotuzumab provides survival benefit to patients with inoperable advanced squamous cell carcinoma of the head and neck: a randomized, open-label, phase IIb, 5-year study in Indian patients. *Oral Oncol,* 50(5), 498–505.

Redmond, D. E. (1970). Tobacco and cancer: the first clinical report, 1761. *N Engl J Med,* 282, 18–23.

Reed, A. L., Califano, J., Cairns, P. et al. (1996). High frequency of p16 (CDKN2/MTS-1/INK4A) inactivation in head and neck squamous cell carcinoma. *Cancer Res,* 56(16), 3630–3633.

Reid, P., Wilson, P., Li, Y. et al. (2017). Current understanding of cancer stem cells: review of their radiobiology and role in head and neck cancers. *Head Neck,* 39(9), 1920–1932.

Reid, P. Experimental investigation of in-vitro cancer stem cells in head and neck cancer cell lines following X-ray irradiation. MSc Thesis, University of South Australia, 2017.

Rettig, E. M., Chung, C. H., Bishop, J. A. et al. (2015). Cleaved NOTCH1 expression pattern in head and neck squamous cell carcinoma is associated with NOTCH1 mutation, HPV status and high-risk features. *Cancer Prev Res,* 8(4), 287–295.

Rey, S., Schito, L., Koritzinsky, M. et al. (2017). Molecular targeting of hypoxia in radiotherapy. *Adv Drug Deliv Rev,* 109, 45–62.

Reya, T., Morrison, S. J., Clarke, M. F. et al. (2001). Stem cells, cancer, and cancer stem cells. *Nature,* 414(6859), 105–111.

Rieckmann, T., Tribius, S., Grob, T. J. et al. (2013). HNSCC cell lines positive for HPV and p16 possess higher cellular radiosensitivity due to an impaired DSB repair capacity. *Radiother Oncol,* 107, 242–246.

Riedel, F., Götte, K., Li, M. et al. (2003). Abrogation of VEGF expression in human head and neck squamous cell carcinoma decreases angiogenic activity in vitro and in vivo. *Int J Oncol,* 23, 577–583.

Riedel, F., Gotte, K., Schwalb, J. et al. (2000). Serum levels of vascular endothelial growth factor in patients with head and neck cancer. *Eur Arch Otorhinolaryngol*, 257, 332–336.

Rijken, P. F., Bernsen, H. J., Peters, J. P. et al. (2000). Spatial relationship between hypoxia and the (perfused) vascular network in a human glioma xenograft: a quantitative multi-parameter analysis. *Int J Radiat Oncol Biol Phys*, 48(2), 571–582.

Rischin, D., Peters, L., Fisher, R. et al. (2005). Tirapazamine, cisplatin, and radiation versus fluo-rouracil, cisplatin, and radiation in patients with locally advanced head and neck cancer: a randomized phase II trial of the Trans-Tasman Radiation Oncology Group (TROG 98.02). *J Clin Oncol*, 23, 79–87.

Rischin, D., Young, R. J., Fisher, R. et al. (2010). Prognostic significance of p16INK4A and human papillomavirus in patients with oropharyngeal cancer treated on TROG 02.02 phase III trial. *J Clin Oncol*, 28(27), 4142–4148.

Ritchie, J. M., Smith, E. M., Summersgill, K. F. et al. (2003). Human papillomavirus infection as a prognostic factor in carcinomas of the oral cavity and oropharynx. *Int J Cancer*, 104(3), 336–344.

Ritchie, K. E. and Nör, J. E. (2013). Perivascular stem cell niche in head and neck cancer. *Cancer Lett*, 338(1), 41–46.

Roccaro, A. M., Hideshima, T., Richardson, P. G. et al. (2006). Bortezomib as an antitumour agent. *Curr Pharm Biotechnol*, 7(6), 441–448.

Roh, K. W., Jang, J. S., Kim, M. S. et al. (2009). Fractionated stereotactic radiotherapy as reirradia-tion for locally recurrent head and neck cancer. *Int J Radiat Oncol Biol Phys*, 74, 1348–1355.

Romesser, P. B., Cahlon, O., Scher, E. D. et al. (2016). Proton beam reirradiation for recurrent head and neck cancer: multi-institutional report on feasibility and early outcomes. *Int J Radiat Oncol Biol Phys*, 95, 386–395.

Rosenberg, B. (1979). Anticancer activity of cis-dichlorodiammineplatinum(II) and some relevant chemistry. *Cancer Treat Rep*, 63(9–10), 1433–1438.

Rosenberg, P. J. (1971). Total laryngectomy and cancer of the larynx. *Arch Otolaryngol*, 94, 313–316.

Rosenquist, K., Wennerberg, J., Schildt, E. B. et al. (2005). Oral status, oral infections and some lifestyle factors as risk factors for oral and oropharyngeal carcinoma. A population-based case-control study in southern Sweden. *Acta Otolaryngol*, 125(12), 1327–1336.

Rothkamm, K. and Löbrich, M. (2003). Evidence for a lack of DNA double-strand break repair in human cells exposed to very low x-ray doses. *Proc Natl Acad Sci USA*, 100, 5057–5062.

Rowland, D. J., Tu, Z., Xu, J. et al. (2006). Synthesis and in vivo evaluation of 2 high-affinity 76Br-labeled sigma2-receptor ligands. *J Nucl Med*, 47(6), 1041–1048.

Rudat, V., Stadler, P., Becker, A. et al. (2001). Predictive value of the tumor oxygenation by means of pO2 histography in patients with advanced head and neck cancer. *Strahlenther Onkol*, 177, 462–468.

Rudat, V., Vanselow, B., Wollensack, P. et al. (2000). Repeatability and prognostic impact of the pretreatment pO(2) histography in patients with advanced head and neck cancer. *Radiother Oncol*, 57, 31–37.

Rudy, S. F., Brenner, J. C., Harris, J. L. et al. (2016). In vivo Wnt pathway inhibition of human squa-mous cell carcinoma growth and metastasis in the chick chorioallantoic model. *J Otolaryngol Head Neck Surg*, 45, 26.

Rwigema, J. C., Choi, J., Lee, N. Y. et al. (2017). Re-irradiation therapy for locally recurrent head and neck cancer: a national survey of practice patterns. *Cancer Invest*, 0, 1–10.

Ryan, J. L. (2012). Ionizing radiation: the good, the bad, and the ugly. *J Invest Dermatol*, 132, 985–993.

Saini, R., Lee, N. V., Liu, K. Y. et al. (2016). Prospects in the application of photodynamic therapy in oral cancer and premalignant lesions. *Cancer*, 8(9), pii:E83.

Salama, J. K., Vokes, E. E., Chmura, S. J. et al. (2006). Long-term outcome of concurrent chemo-therapy and reirradiation for recurrent and second primary head-and-neck squamous cell carcinoma. *Int J Radiat Oncol Biol Phys*, 64, 382–391.

Santini, V. (2001). Amifostine: chemotherapeutic and radiotherapeutic protective effects. *Expert Opinion Pharmacother*, 2, 479–489.

Sato, H., Hatori, M., Ando, Y. et al. (2009). S-1 mediates the inhibition of lymph node metastasis in oral cancer cells. *Oncol Rep*, 22(4), 719–724.

Sato, J., Kitagawa, Y., Yamazaki, Y. et al. (2013). 18F-fluoromisonidazole PET uptake is correlated with hypoxia-inducible factor-1alpha expression in oral squamous cell carcinoma. *J Nucl Med*, 54, 1060–1065.

Sato, J. D., Kawamoto, T., Le, A. D. et al. (1983). Biological effects in vitro of monoclonal antibodies to human epidermal growth factor receptors. *Mol Biol Med*, 1, 511–529.

Saunders, M. I., Rojas, A. M., Parmar, M. K. et al. (2010). CHART Trial Collaborators. Mature results of a randomized trial of accelerated hyperfractionated versus conventional radiotherapy in head-and-neck cancer. *Int J Radiat Oncol Biol Phys*, 77(1), 3–8.

Schantz, S. P., Hsu, T. C., Ainslie, N. et al. (1989). Young adults with head and neck cancer express increased susceptibility to mutagen-induced chromosome damage. *JAMA*, 262(23), 3313–3315.

Schantz, S. P., Zhang, Z. F., Spitz, M. S. et al. (1997). Genetic susceptibility to head and neck cancer: Interaction between nutrition and mutagen sensitivity. *Laryngoscope*, 107(6), 765–781.

Schneider, S., Thurnher, D., Kloimstein, P. et al. (2011). Expression of the Sonic hedgehog pathway in squamous cell carcinoma of the skin and the mucosa of the head and neck. *Head Neck*, 33(2), 244–250.

Schrijvers, M. L., van der Laan, B. F., de Bock, G. H. et al. (2008). Overexpression of intrinsic hypoxia markers HIF1alpha and CA-IX predict for local recurrence in stage T1-T2 glottic laryngeal carcinoma treated with radiotherapy. *Int J Radiat Oncol Biol Phys*, 72(1), 161–169.

Schuller, D. E., Ozer, E., Agrawal, A. et al. (2007). Multimodal intensification regimens for advanced, resectable, previously untreated squamous cell cancer of the oral cavity, oropharynx, or hypopharynx: a 12-year experience. *Arch Otolaryngol Head Neck Surg*, 133, 320–326.

Schwartz, S. M., Daling, J. R., Doody, D. R. et al. (1998). Oral cancer risk in relation to sexual history and evidence of human papillomavirus infection. *J Natl Cancer Inst*, 90(21), 1626–1636.

Schwarz, G. (1909). Ueber Desensibilisierung gegen rontgen- und radiumstrahlen. *Munchener Medizinische Wochenschrift*, 24, 1–2.

Sedlacek, H. H. (2000). Kinase inhibitors in cancer therapy. A look ahead. *Drugs*, 59, 435–476.

Seifert, E., Schadel, A., Haberkorn, U. et al. (1992). Evaluating the effectiveness of chemotherapy in patients with head-neck tumors using positron emission tomography (PET scan). *Head Neck Oncol*, 40, 90–93.

Seiwert, T. Y., Burtness, B., Mehra, R. et al. (2016). Safety and clinical activity of pembrolizumab for treatment of recurrent or metastatic squamous cell carcinoma of the head and neck (KEYNOTE-012): an open-label, multicentre, phase 1b trial. *Lancet Oncol*, 17(7), 956–965.

Senger, D. R., Galli, S. J., Dvorak, A. M. et al. (1983). Tumor cells secrete a vascular permeability factor that promotes accumulation of ascites fluid. *Science*, 219(4587), 983–985.

Seol, Y. M., Kwon, B. R., Song, M. K. et al. (2010). Measurement of tumor volume by PET to evaluate prognosis in patients with head and neck cancer treated by chemoradiation therapy. *Acta Oncol*, 49, 201–208.

Shackleton, M., Quintana, E., Fearon, E. R. et al. (2009). Heterogeneity in cancer: cancer stem cells versus clonal evolution. *Cell*, 138(5), 822–829.

Shikama, N., Kumazaki, Y., Tsukamoto, N. et al. (2013). Validation of nomogram-based prediction of survival probability after salvage reirradiation of head and neck cancer. *Jpn J Clin Oncol*, 43, 154–160.

Shirinian, M. H., Weber, R. S., Lippman, S. M. et al. (1994). Laryngeal preservation by induction chemotherapy plus radiotherapy in locally advanced head and neck cancer: The M.D. Anderson Cancer Center experience. *Head Neck*, 16, 39–44.

Shrivastava, S., Steele, R., Sowadski, M. et al. (2015). Identification of molecular signature of head and neck cancer stem-like cells. *Sci Rep*, 5, 7819.

Siegel, R. L., Miller, K. D., and Jemal, A. (2015). Cancer statistics, 2015. *CA Cancer J Clin*, 65(1), 5–29.

Singhi, A. D. and Westra, W. H. (2010). Comparison of human papillomavirus in situ hybridization and p16 immunohistochemistry in the detection of human papillomavirus-associated head and neck cancer based on a prospective clinical experience. *Cancer*, 116, 2166–2173.

Siu, L. L., Waldron, J. N., Chen, B. E. et al. (2016). Effect of standard radiotherapy with cisplatin vs accelerated radiotherapy with panitumumab in locoregionally advanced squamous cell head and neck carcinoma: a randomized clinical trial. *JAMA Oncol*, doi: 10.1001/jamaoncol.2016.4510. [Epub ahead of print].

Skladowski, B., Maciejewski, M., Golen, R. et al. (2006). Continuous accelerated 7-days-a-week radiotherapy for head-and-neck cancer: long-term results of phase III clinical trial. *Int J Radiat Oncol Biol Phys*, 66, 706–713.

Skladowski, K., Maciejewski, B., Golen, M. et al. (2000). Randomized clinical trial on 7-day-continuous accelerated irradiation (CAIR) of head and neck cancer—report on 3-year tumour control and normal tissue toxicity. *Radiother Oncol*, 55, 101–110.

Smith, B. D., Smith, G. L., Carter, D. et al. (2000). Prognostic significance of vascular endothelial growth factor protein levels in oral and oropharyngeal squamous cell carcinoma. *J Clin Oncol*, 18, 2046–2052.

Smith, E. M., Ritchie, J. M., Summersgill, K. F. et al. (2004). Age, sexual behaviour and human papillomavirus infection in oral cavity and oropharyngeal cancers. *Int J Cancer*, 108(5), 766–772.

Snijders, P. J. (1992). Prevalence and expression of human papillomavirus in tonsillar carcinomas indicating a possible viral etiology. *Int J Cancer*, 51(6), 845–850.

Solares, C. A., Fritz, M. A., and Esclamado, R. M. (2005). Oncologic effectiveness of selective neck dissection in the N0 irradiated neck. *Head Neck*, 27, 415–420.

Sørensen, B. S., Busk, M., Horsman, M. R. et al. (2014). Effect of radiation on cell proliferation and tumour hypoxia in HPV-positive head and neck cancer in vivo models. *Anticancer Res*, 34(11), 6297–6304.

Sørensen, B. S., Busk, M., Olthof, N. et al. (2013). Radiosensitivity and effect of hypoxia in HPV positive head and neck cancer cells. *Radiother Oncol*, 108(3), 500–505.

Soule, B. P., Hyodo, F., Matsumoto, K. et al. (2007). Therapeutic and clinical applications of nitroxide compounds. *Antioxid Redox Signal*, 9(10), 1731–1743.

Souvatzoglou, M., Grosu, A. L., Röper, B. et al. (2007). Tumour hypoxia imaging with [^{18}F]FAZA PET in head and neck cancer patients: a pilot study. *Eur J Nucl Med Mol Imaging*, 34, 1566–1575.

Specenier, P. and Vermorken, J. B. (2013). Cetuximab: its unique place in head and neck cancer treatment. *Biologics*, 7, 77–90.

Spencer, S. A., Harris, J., Wheeler, R. H. et al. (2001). RTOG 96-10: reirradiation with concurrent hydroxyurea and 5-fluorouracil in patients with squamous cell cancer of the head and neck. *Int J Radiat Oncol Biol Phys*, 51, 1299–1304.

Spencer, S. A., Wheeler, R. H., Peters, G. E. et al. (1999). Concomitant chemotherapy and reirradiation as management for recurrent cancer of the head and neck. *Am J Clin Oncol*, 22, 1–5.

Spitz M. R., Fueger J. J., Beddingfield N. A. et al. (1989). Chromosome sensitivity to bleomycin-induced mutagenesis, an independent risk factor for upper aerodigestive tract cancers. *Cancer Res*, 49(16), 4626–4628.

Sprong, D., Janssen, H. L., Vens, C. et al. (2006). Resistance of hypoxic cells to ionizing radiation is influenced by homologous recombination status. *Int J Radiat Oncol Biol Phys*, 64, 562–572.

Stadler, P., Becker, A., Feldmann, H. J. et al. (1999). Influence of the hypoxic subvolume on the survival of patients with head and neck cancer. *Int J Radiat Oncol Biol Phys*, 44, 749–754.

Steel, G. G. (1968). Cell loss from experimental tumours. *Cell Proliferation*, 1, 193–207.

Steel, G. G., McMillan, T. J., and Peacock, J. H. (1989). The 5 Rs of radiotherapy. *Int J Radiat Biol*, 56, 1045–1048.

Stordal, B., Pavlakis, N., and Davey, R. (2007). Oxaliplatin for the treatment of cisplatin-resistant cancer: a systematic review. *Cancer Treat Rev*, 33(4), 347–357.

Strandqvist, M. (1944). Studien über die cumulative Wirkung der Röntgenstrahlen bei Fraktionierung. *Acta Radiol*, 55(Suppl.), 1–300.

Stransky, N., Egloff, A. M., Tward, A. D. et al. (2011). The mutational landscape of head and neck squamous cell carcinoma. *Science*, 333, 1157–1160.

Strati, K., Pitot, H. C., and Lambert, P. F. (2006). Identification of biomarkers that distinguish human papillomavirus (HPV)-positive versus HPV-negative head and neck cancers in a mouse model. *Proc Natl Acad Sci USA*, 103(38), 14152–14157.

Strnad, V., Geiger, M., and Lotter, M. (2003). The role of pulsed-dose-rate brachytherapy in previously irradiated head-and-neck cancer. *Brachytherapy*, 2, 158–163.

Studer, G., Peponi, E., Kloeck, S. et al. (2010). Surviving hypopharynx-larynx carcinoma in the era of IMRT. *Int J Radiat Oncol Biol Phys*, 77, 1391–1396.

Sturgis, E. M. and Cinciripini, P. M. (2007). Trends in head and neck cancer incidence in relation to smoking prevalence: an emerging epidemic of human papillomavirus-associated cancers? *Cancer*, 11(7), 1429–1435.

Sturgis, E. M., Moore, B. A., Glisson, B. S. et al. (2005). Neoadjuvant chemotherapy for squamous cell carcinoma of the oral tongue in young adults: a case series. *Head Neck*, 27, 748–756.

Sulman, E. P., Schwartz, D. L., Le, T. T. et al. (2009). IMRT reirradiation of head and neck cancer-disease control and morbidity outcomes. *Int J Radiat Oncol Biol Phys*, 73, 399–409.

Sun, A., Johansson, S., Turesson, I. et al. (2012). Imaging tumor perfusion and oxidative metabolism in patients with head-and-neck cancer using 1-[(11)C]-acetate PET during radiotherapy: preliminary results. *Int J Radiat Oncol Biol Phys*, 82(2), 554–560.

Sun, W., Gaykalova, D. A., Ochs, M. F. et al. (2014). Activation of the NOTCH pathway in head and neck cancer. *Cancer Res*, 74(4), 1091–1104.

Sureka, C. S. and Armpilia, C. Radiation Biology for Medical Physicists. CRC Press; 2017.

Symonds, R. P., McIlroy, P., Khorrami, J. et al. (1996). The reduction of radiation mucositis by selective decontamination antibiotic pastilles: a placebo-controlled double-blind trial. *Br J Cancer*, 74, 312–317.

Syrjänen, K. J., Pyrhönen, S., Syrjänen, S. M. et al. (1983). Immunohistochemical demonstration of human papilloma virus (HPV) antigens in oral squamous cell lesions. *Br J Oral Surg*, 21(2), 147–153.

Takiar, V., Garden, A. S., Ma, D. et al. (2016). Reirradiation of head and neck cancers with intensity modulated radiation therapy: outcomes and analyses. *Int J Radiat Oncol Biol Phys*, 95(4), 1117–1131.

Talamini, R., Bosetti, C., La Vecchia, C. et al. (2012). Combined effect of tobacco and alcohol on laryngeal cancer risk: a case-control study. *Cancer Causes Control*, 13(10), 1367–1375.

Tamatani, T., Takamaru, N., Hara, K. et al. (2013). Bortezomib-enhanced radiosensitization through the suppression of radiation-induced nuclear factor-kappa B activity in human oral cancer cells. *Int J Oncol*, 42, 935–944.

Tang, A. L., Hauff, S. J., Owen, J. H. et al. (2012). UM-SCC-104: a new human papillomavirus-16-positive cancer stem cell-containing head and neck squamous cell carcinoma cell line. *Head Neck*, 34(10), 1480–1491.

Tang, A. L., Owen, J. H., Hauff, S. J. et al. (2013). Head and neck cancer stem cells: the effect of HPV—an in vitro and mouse study. *Otolaryngol Head Neck Surg*, 149(2), 252–260.

Tanvetyanon, T., Padhya, T., McCaffrey, J. et al. (2009). Prognostic factors for survival after salvage reirradiation of head and neck cancer. *JCO*, 27, 1983–1991.

Tarvainen, L., Suojanen, J., and Kyyronen, P. (2017). Occupational risk for oral cancer in onrdic countries. *Anticancer Res*, 37(6), 3221–3228.

Taussky, D., Dulguerov, P., and Allal, A. S. (2005). Salvage surgery after radical accelerated radiotherapy with concomitant boost technique for head and neck carcinomas. *Head Neck,* 27, 182–186.

Tesselaar, E., Flejmer, A. M., Farnebo, S. et al. (2017). Changes in skin microcirculation during radiation therapy for breast cancer. *Acta Oncol,* 56, 1072–1080.

Thames, H. D. (1985). An 'incomplete-repair' model for survival after fractionated and continuous irradiations. *Int J Radiat Biol,* 47, 319–339.

Thames, H. D. Jr. (1988). Early fractionation methods and the origins of the NSD concept. *Acta Oncol,* 27(2), 89–103.

Thames, H. D. and Hendry, J. H. Fractionation in Radiotherapy. London: Taylor & Francis; 1987.

Thames, H. D., Withers, H. R., Peters, L. J. et al. (1982). Changes in early and late radiation responses with altered dose fractionation: implications for dose-survival relationships. *Int J Radiat Oncol Biol Phys,* 8(2), 219–226.

The Quarterback Trial: Reduced Dose Radiotherapy for HPV+ Oropharynx Cancer. Available from https://clinicaltrials.gov/ct2/show/NCT01706939. (Accessed on 16/01/2017.)

The Royal College of Radiologists (RCR). The Timely Delivery of Radical Radiotherapy: Standards and Guidelines for the Management of Unscheduled Treatment Interruptions, 3rd edition. London: Royal College of Radiologists; 2008.

Thomas, B. and Devi, P. U. (1987). Chromosome protection by WR-2721 and MPG-single and combination treatments. *Strahlenther Onkol,* 163, 807–810.

Thomlinson, R. H. and Gray, L. H. (1955). The histological structure of some human lung cancers and the possible implications for radiotherapy. *Br J Cancer,* 9, 539–549.

Thorwarth, D., Eschmann, S. M., Paulsen, F. et al. (2007). Hypoxia dose painting by numbers: a planning study. *Int J Radiat Oncol Biol Phys,* 68, 291–300.

Thumfart, W., Weidenbecher, M., Waller, G. et al. (1978). Chronic mechanical trauma in the aetiology of oro-pharyngeal carcinoma. *J Maxillofac Surg,* 6(3), 217–221.

Toma-Dasu, I. and Dasu, A. (2015). Towards multidimensional radiotherapy: key challenges for treatment individualization. *Comput Mathem Meth Medicine,* 934380.

Toma-Dasu, I., Uhrdin, J., Antonovic, L. et al. (2012). Dose prescription and treatment planning based on FMISO-PET hypoxia. *Acta Oncol,* 51, 222–230.

Toma-Dasu, I., Uhrdin, J., Dasu, A. et al. (2009). Therapy optimization based on non-linear uptake of PET tracers versus "linear dose painting". *IFMBE Proceedings,* 25/I, 221–224.

Toma-Dasu, I., Uhrdin J., Lazzeroni, M. et al. (2015). Evaluating tumor response of non-small cell lung cancer patients with 18F-fludeoxyglucose positron emission tomography: potential for treatment individualization. *Int J Radiat Oncol Biol Phys,* 91(2), 376–384.

Toma-Dasu, I., Waites, A., Dasu, A. et al. (2001). Theoretical simulation of oxygen tension measurement in tissues using a microelectrode: I. The response function of the electrode. *Physiol Meas,* 22, 713–725.

Torre, L. A., Bray, F., Siegel, R. L. et al. (2015). Global cancer statistics, 2012. *CA Cancer J Clin,* 65(2), 87–108.

Tortochaux, J., Tao, Y., Tournay, E. et al. (2011). Randomized phase III trial (GORTEC98-03) comparing re-irradiation plus chemotherapy versus methotrexate in patients with recurrent or a second primary head and neck squamous cell carcinoma, treated with a palliative intent. *Radiother Oncol,* 100, 70–75.

Toulany, M., Dittmann, K., Krüger, M. et al. (2005). Radioresistance of K-Ras mutated human tumor cells is mediated through EGFR-dependent activation of PI3K-AKT pathway. *Radiother Oncol,* 76, 143–150.

Toulany, M., Kasten-Pisula, U., Brammer, I. et al. (2006). Blockage of epidermal growth factor receptor-phosphatidylinositol 3-kinase-Akt signaling increases radiosensitivity of K-RAS mutated human tumor cells in vitro by affecting DNA repair. *Clin Cancer Res,* 12, 4119–4126.

Tran, L. B., Bol, A., Labar, D. et al. (2012). Hypoxia imaging with the nitroimidazole 18F-FAZA PET tracer: a comparison with OxyLite, EPR oximetry and 19F-MRI relaxometry. *Radiother Oncol*, 105, 29–35.

Tran, N., McLean, T., Zhang, X. et al. (2007). MicroRNA expression profiles in head and neck cancer cell lines. *Biochem Biophys Res Commun*, 358, 12–17.

Trapasso, S. and Allegra, E. (2012). Role of CD44 as a marker of cancer stem cells in head and neck cancer. *Biologics*, 6, 379–383.

Travis, E. L. and Tucker, S. L. (1987). Isoeffect models and fractionated radiation therapy. *Int J Radiat Oncol Biol Phys*, 13, 283–287.

Trinkaus, M. E., Hicks, R. J., Young, R. J. et al. (2014). Correlation of p16 status, hypoxic imaging using [18F]-misonidazole positron emission tomography and outcome in patients with loco-regionally advanced head and neck cancer. *J Med Imaging Radiat Oncol*, 58(1), 89–97.

TROG, Weekly Cetuximab/RT Vs Weekly Cisplatin/RT in HPV-Associated Oropharyngeal Squamous Cell Carcinoma (HPVOropharynx). Available from http://www.trog.com.au /TROG-1201-HPV. (Accessed on 16/01/2017.)

Troost, E. G., Bussink, J., Hoffmann, A. L. et al. (2010). 18F-FLT PET/CT for early response monitoring and dose escalation in oropharyngeal tumors. *J Nucl Med*, 51(6), 866–874.

Trott, K. (1999). The mechanisms of acceleration of repopulation in squamous epithelia during daily irradiation. *Acta Oncol*, 38, 153–157.

Trott, K. and Kummermehr, J. (1991). Accelerated repopulation in tumours and normal tissues. *Radiother Oncol*, 22, 159–160.

Trotti, A. (2000). Toxicity in head and neck cancer: a review of trends and issues. *Int J Radiat Oncol Biol Phys*, 47(1), 1–12.

Trotti, A., Byhardt, R., Stetzm J. et al. (2000). Common toxicity criteria: version 2.0. An improved reference for grading the acute effects of cancer treatment: impact on radiotherapy. *Int J Radiat Oncol Biol Phys*, 47(1), 13–47.

Trotti, A., Colevas, A. D., Setser, A. et al. (2003). CTCAE v3.0: development of a comprehensive grading system for the adverse effects of cancer treatment. *Semin Radiat Oncol*, 13(3), 176–181.

Troy, J. D., Weissfeld, J. L., Youk, A. O. et al. (2013). Expression of EGFR, VEGF, and NOTCH1 suggest differences in tumor angiogenesis in HPV–positive and HPV-negative head and neck squamous cell carcinoma. *Head Neck Pathol*, 7(4), 344–355.

Truong, M. T., Zhang, Q., Rosenthal, D. I. et al. (2017). Quality of life and performance status from a substudy conducted within a prospective phase 3 randomized trial of concurrent accelerated radiation plus cisplatin with or without cetuximab for locally advanced head and neck carcinoma: NRG Oncology Radiation Therapy Oncology Group 0522. *Int J Radiat Oncol Biol Phys*, 97(4), 687–699.

Tse, G. M., Chan, A. W., Yu, K. H. et al. (2007). Strong immunohistochemical expression of vascular endothelial growth factor predicts overall survival in head and neck squamous cell carcinoma. *Ann Surg Oncol*, 14, 3558–3565.

Tsujii, H., Kamada, T., Baba, M. et al. (2008). Clinical advantages of carbonion radiotherapy. *New J Phys*, 10, 075009.

Tu, Z., Xu, J., Jones, L. A. et al. (2007). Fluorine-18-labeled benzamide analogues for imaging the σ2 receptor status of solid tumors with positron emission tomography. *J Med Chem*, 50, 3194–3204.

Tubiana, M. (1988). Repopulation in human tumours. *Acta Oncol*, 27, 83–88.

Turesson, I., Carlsson, J., Brahme, A. et al. (2003). Biological response to radiation therapy. *Acta Oncol*, 42, 92–106.

Turesson, I., Nyman, J., Holmberg, E. et al. (1996). Prognostic factors for acute and late skin reactions in radiotherapy patients. *Int J Radiat Oncol Biol Phys*, 36, 1065–1075.

Ukpo, O. C., Flanagan, J. J., Ma, X. J. et al. (2011). High-risk human papillomavirus E6/E7 mRNA detection by a novel in situ hybridization assay strongly correlates with p16 expression and patient outcomes in oropharyngeal squamous cell carcinoma. *Am J Surg Pathol*, 35, 1343–1350.

Um, J. H., Kang, C. D., Bae, J. H. et al. (2004). Association of DNA-dependent protein kinase with hypoxia inducible factor-1 and its implication in resistance to anticancer drugs in hypoxic tumor cells. *Exp Mol Med,* 36, 233–242.

Unger, K. R., Lominska, C. E., Deeken, J. F. et al. (2010). Fractionated stereotactic radiosurgery for reirradiation of head-and-neck cancer. *Int J Radiat Oncol Biol Phys,* 77, 1411–1419.

Urban, D., Corry, J., and Rischin, D. (2014). What is the best treatment for patients with human papillomavirus-positive and -negative oropharyngeal cancer? *Cancer,* 120(10), 1462–1470.

van Beek, K. M., Kaanders, J. H., Janssens, G. O. et al. (2016). Effectiveness and toxicity of hypo-fractionated high-dose intensity-modulated radiotherapy versus 2- and 3-dimensional radiotherapy in incurable head and neck cancer. *Head Neck,* 38(Suppl 1), E1264–1270.

Van der Schueren, E., Van den Bogaert, W., Vanuytsel, L. et al. (1990). Radiotherapy by multiple fractions per day (MFD) in head and neck cancer: acute reactions of skin and mucosa. *Int J Radiat Oncol Biol Phys,* 19, 301–311.

van Luijk, P., Pringle, S., Deasy, J. O. et al. (2015). Sparing the region of the salivary gland containing stem cells preserves saliva production after radiotherapy for head and neck cancer. *Sci Transl Med,* 7, 305ra147.

van Waes, C., Chang, A. A., Lebowitz, P. F. et al. (2005). Inhibition of nuclear factor-kappa B and target genes during combined therapy with proteasome inhibitor bortezomib and reirradiation in patients with recurrent head-and-neck squamous cell carcinoma. *Int J Rad Oncol Biol Phys,* 63(5), 1400–1412.

Vargo, J. R., Ward, M. C., Caudell, J. J. et al. (2017). A multi-institutional comparison of SBRT and IMRT for definitive reirradiation of recurrent or second primary head and neck cancer. *Int J Radiat Oncol Biol Phys,* (in press).

Varia, M. A., Calkins-Adams, D. P., Rinker, L. H. et al. (1998). Pimonidazole: a novel hypoxia marker for complementary study of tumor hypoxia and cell proliferation in cervical carcinoma. *Gynecol Oncol,* 71, 270–277.

Vaupel, P. (1997). Blood flow and oxygenation status of head and neck carcinomas. *Adv Experim Med Biol,* 428, 89–95.

Vaupel, P., Kallinowski, F., and Okunieff, P. (1989). Blood flow, oxygen and nutrient supply, and metabolic microenvironment of human tumors: a review. *Cancer Research,* 49(23), 6449–6465.

Vaupel, P., Schlenger, K., Knoop, C. et al. (1991). Oxygenation of human tumors: evaluation of tissue oxygen distribution in breast cancers by computerized O_2 tension measurements. *Cancer Research,* 51, 3316–3322.

Venuti, A. and Paolini, F. (2012). HPV detection methods in head and neck cancer. *Head Neck Pathol,* 6(Suppl 1), 63–74.

Vermorken, J. B., Mesia, R., Rivera, F. et al. (2008). Platinum-based chemotherapy plus cetuximab in head and neck cancer. *N Engl J Med,* 359(11), 1116–1127.

Vermorken, J. B., Psyrri, A., Mesia, R. et al. (2014). Impact of tumor HPV status on outcome in patients with recurrent and/or metastatic squamous cell carcinoma of the head and neck receiving chemotherapy with or without cetuximab: retrospective analysis of the phase III EXTREME trial. *Ann Oncol,* 5(4), 801–807.

Vermorken, J. B., Stöhlmacher-Williams, J., Davidenko, I. et al. (2013). Cisplatin and fluorouracil with or without panitumumab in patients with recurrent or metastatic squamous-cell carcinoma of the head and neck (SPECTRUM): an open-label phase 3 randomised trial. *Lancet Oncol,* 14(8), 697–710.

Vlashi, E., Chen, A. M., Boyrie, S. et al. (2016). Radiation-induced dedifferentiation of head and neck cancer cells into cancer stem cells depends on human papillomavirus status. *Int J Radiat Oncol Biol Phys,* 94(5), 1198–1206.

Vlashi, E., McBride, W. H., and Pajonk, F. (2009). Radiation responses of cancer stem cells. *J Cell Biochem,* 108(2), 339–342.

Vojtisek, R., Ferda, J., and Finek, J. (2015). Effectiveness of PET/CT with (18)F-fluorothymidine in the staging of patients with squamous cell head and neck carcinomas before radiotherapy. *Rep Pract Oncol Radiother,* 20(3), 210–216.

Wahl, R. L., Jacene, H., Kasamon, Y. et al. (2009). From RECIST to PERCIST: evolving considerations for PET response criteria in solid tumors. *J Nucl Med,* 50(Suppl 1), 122S–150S.

Walpole, S. C., Prieto-Merino, D., Edwards, P. et al. (2012). The weight of nations: an estimation of adult human biomass. *BMC Public Health,* 12(1), 439.

Wang, C. J. and Knecht, R. (2011). Current concepts of organ preservation in head and neck cancer. *Eur Arch Otorhinolaryngol,* 268(4), 481–487.

Wang, K., Heron, D. E., Flickinger, J. C. et al. (2012). A retrospective, deformable registration analysis of the impact of PET-CT planning on patterns of failure in stereotactic body radiation therapy for recurrent head and neck cancer. *Head Neck Oncol,* 4, 12.

Wang, M. B., Liu, I. Y., Gornbein, J. A., and Nguyen, C. T. (2015). HPV-positive oropharyngeal carcinoma: a systematic review of treatment and prognosis. *Otolaryngol Head Neck Surg,* 153, 758–769.

Wang, Q., Wu, Z., Han, D. et al. (1999). Growth inhibition of human laryngeal cancer cell with the adenovirus-mediated p53 gene. *Chinese J Cancer Res,* 11(3), 157–160.

Wang, Y., Dong, L., Bi, Q. et al. (2010). Investigation of the efficacy of a bevacizumab-cetuximab-cisplatin regimen in treating head and neck squamous cell carcinoma in mice. *Target Oncol,* 5(4), 237–243.

Wang, Z., Valera, J. C., Zhao, X. et al. (2017). mTOR co-targeting strategies for head and neck cancer therapy. *Cancer Metastasis Rev,* 36, 491–502.

Ward, M. J., Thirdborough, S. M., Mellows, T. et al. (2014). Tumour-infiltrating lymphocytes predict outcome in HPV-positive oropharyngeal cancer. *Br J Cancer,* 110, 489–500.

Wasserman, J. K., Rourke, R., Purgina, B. et al. (2017). HPV DNA in saliva from patients with SCC of the head and neck is specific for p16-positive oropharyngeal tumours. *J Otolaryngol Head Neck Surg,* 46(1), 3.

Weber, C. N., Cerniglia, G. J., Maity, A. et al. (2007). Bortezomib sensitizes head and neck carcinoma cells SQOB to radiation. *Cancer Biol Ther,* 6, 156–159.

Werness, B. A., Levine, A. J., and Howley, P. M. (1990). Association of human papillomavirus types 16 and 18 E6 proteins with p53. *Science,* 248(4951), 76–79.

West, C. M., Elliott, R. M., and Burnet, N. G. (2007). The genomics revolution and radiotherapy. *Clin Oncol,* 19, 470–480.

Westra, W. H. (2014). Detection of human papillomavirus (HPV) in clinical samples: evolving methods and strategies for the accurate determination of HPV status of head and neck carcinomas. *Oral Oncol,* 50, 771–779.

Wheeler, K. T., Wang, L. M., Wallen, C. A. et al. (2000). Sigma-2 receptors as a biomarker of proliferation in solid tumours. *Br J Cancer,* 82(6), 1223–1232.

Wick, C. C., Rezaee, R. P., Wang, T. et al. (2017). Use of concurrent chemoradiation in advanced staged (T4) laryngeal cancer. *Am J Otolaryngol,* 38(1), 72–76.

Wiest, T., Schwarz, E., Enders, C. et al. (2002). Involvement of intact HPV16 E6/E7 gene expression in head and neck cancers with unaltered p53 status and perturbed pRb cell cycle control. *Oncogene,* 21(10), 1510–1517.

Williams, M. V., Denekamp, J., and Fowler, J. F. (1985). A review of α/β values for experimental tumours: implications for clinical studies of altered fractionation. *Int J Radiat Oncol Biol Phys,* 11, 87–96.

Williamson, S. K., Moon, J., Huang, C. H. et al. (2010). Phase II evaluation of sorafenib in advanced and metastatic squamous cell carcinoma of the head and neck: Southwest Oncology Group Study S0420. *J Clin Oncol,* 28(20), 3330–3335.

Wilson, R. (2010). Radiological use of fast protons. *Radiology,* 47, 487–491.

Withers, H. R. and Elkind, M. M. (1969). Radiosensitivity and fractionation response of crypt cells of mouse jejunum. *Radiat Res,* 38, 598–613.

Withers, H. R., Taylor, J. M., and Maciejewski, B. (1988). The hazard of accelerated tumor clonogen repopulation during radiotherapy. *Acta Oncol,* 27(2), 131–146.

Withers, H. R., Taylor, J. M., and Maciejewski, B. (1988). Treatment volume and tissue tolerance. *Int J Radiat Oncol Biol Phys,* 14, 751–759.

Withers, H. R. The four R's of radiotherapy. In: Lett, J. T., Alder, H. (Eds.). Advances in Radiation Biology. New York: Academic Press; 1975. pp. 241–271.

Withers, H. R. (1993). Treatment-induced accelerated human tumor growth. *Semin Radiat Oncol,* 3, 135–143.

Wouters, B. G., van den, B. T., Magagnin, M. G. et al. (2005). Control of the hypoxic response through regulation of mRNA translation. *Semin Cell Dev Biol,* 16, 487–501.

Xu, J., He, X., Cheng, K. et al. (2014). Concurrent chemoradiotherapy with nedaplatin plus paclitaxel or fluorouracil for locoregionally advanced nasopharyngeal carcinoma: survival and toxicity. *Head Neck,* 36, 1474–1480.

Xu, Q., Wang, C., Yuan, X. et al. (2017). Prognostic value of tumor-infiltrating lymphocytes for patients with head and neck squamous cell carcinoma. *Trans Oncol,* 10(1), 10–16.

Xue, X., You, S., Zhang, Q. et al. (2012). Mitaplatin increases sensitivity of tumor cells to cisplatin by inducing mitochondrial dysfunction. *Molec Pharma,* 9(3), 634–644.

Yamaoka, T., Ohba, M., and Ohmori, T. (2017). Molecular-targeted therapies for epidermal growth factor receptor and its resistance mechanisms. *Int J Mol Sci,* 18(11), pii:E2420.

Yan, M., Zhang, Y., He, B. et al. (2014). IKKα restoration via EZH2 suppression induces nasopharyngeal carcinoma differentiation. *Nat Commun,* 5, 3661.

Yanamoto, S., Yamada, S., Takahashi, H. et al. (2014). Expression of the cancer stem cell markers CD44v6 and ABCG2 in tongue cancer: effect of neoadjuvant chemotherapy on local recurrence. *Int J Oncol,* 44(4), 1153–1162.

Yang, M. H., Chiang, W. C., Chou, T. Y. et al. (2006). Increased NBS1 expression is a marker of aggressive head and neck cancer and overexpression of NBS1 contributes to transformation. *Clin Cancer Res,* 12(2), 507–515.

Yao, M., Galanopoulos, N., Lavertu, P. et al. (2015). Phase II study of bevacizumab in combination with docetaxel and radiation in locally advanced squamous cell carcinoma of the head and neck. *Head Neck,* 37(11), 1665–1671.

Yin, L., Bian, X. H., Wang, X. et al. (2015). Long-term results of concurrent chemoradiotherapy for advanced N2-3 stage nasopharyngeal carcinoma. *PLoS One,* 10(9), e0137383.

Yonesaka, K., Zejnullahu, K., Okamoto, I. et al. (2011). Activation of ERBB2 signaling causes resistance to the EGFR-directed therapeutic antibody cetuximab. *Sci Transl Med,* 3, 99ra86.

Yoshikawa, A., Saura, R., and Matsubara, T. (1997). A mechanism of cisplatin action: antineoplastic effect through inhibition of neovascularization. *Kobe J Med Sci,* 43, 109–120.

Young, R. J., Rischin, D., Fisher, R. et al. (2011). Relationship between epidermal growth factor receptor status, p16(INK4A), and outcome in head and neck squamous cell carcinoma. *Cancer Epidemiol Biomarkers Prev,* 20(6), 1230–1237.

Young, W. K., Vojnovic, B., and Wardman, P. (1996). Measurement of oxygen tension in tumours by time-resolved fluorescence. *Br J Cancer,* Suppl 27, S256–S259.

Yüce, I., Bayram, A., Cağlı, S. et al. (2011). The role of CD44 and matrix metalloproteinase-9 expression in predicting neck metastasis of supraglottic laryngeal carcinoma. *Am J Otolaryngol,* 32, 141–146.

Yunhong, T., Jie, L., Yunming, T. et al. (2017). Efficacy and safety of anti-EGFR agents administered concurrently with standard therapies for patients with head and neck squamous cell carcinoma: a systematic review and meta-analysis of randomized controlled trials. *Int J Cancer,* doi: 10.1002/ijc.31157.

Zackrisson, B., Mercke, C., Strander, H. et al. (2003). A systematic overview of radiation therapy effects in head and neck cancer. *Acta Oncol,* 42, 443–461.

Zeman, E. M., Brown, J. M., Lemmon, M. J. et al. (1986). SR-4233: a new bioreductive agent with high selective toxicity for hypoxic mammalian cells. *Int J Radiat Oncol Biol Phys,* 12, 1239–1242.

Zhang, M., Kumar, B., Piao, L. et al. (2014). Elevated intrinsic cancer stem cell population in human papillomavirus-associated head and neck squamous cell carcinoma. *Cancer,* 120(7), 992–1001.

Zhang, S. W., Xiao, S. W., Liu, C. Q. et al. (2005). Recombinant adenovirus-p53 gene therapy combined with radiotherapy for head and neck squamous-cell carcinoma. *Zhonghua Zhong Liu Za Zhi,* 27(7), 426–428.

Zhang, Z. F., Morgenstern, H., Spitz, M. R. et al. (2000). Environmental tobacco smoking, mutagen sensitivity, and head and neck squamous cell carcinoma. *Cancer Epidemiol Biomarkers Prev,* 9(10), 1043–1049.

Zhou, S., Schuetz, J. D., Bunting, K. D. et al. (2001). The ABC transporter Bcrp1/ABCG2 is expressed in a wide variety of stem cells and is a molecular determinant of the side-population phenotype. *Nat Med,* 67, 1028–1034.

Ziemann, F., Arenz, A., Preising, S. et al. (2015). Increased sensitivity of HPV-positive head and neck cancer cell lines to x-irradiation ± Cisplatin due to decreased expression of E6 and E7 oncoproteins and enhanced apoptosis. *Am J Cancer Res,* 5(3), 1017–1031.

Zips, D., Zophel, K., Abolmaali, N. et al. (2012). Exploratory prospective trial of hypoxia-specific PET imaging during radiochemotherapy in patients with locally advanced head-and-neck cancer. *Radiother Oncol,* 105, 21–28.

Zöller, M. (2011). CD44: can a cancer-initiating cell profit from an abundantly expressed molecule? *Nat Rev Cancer,* 11(4), 254–267.

Zolzer, F., Streffer, C. (2002). Increased radiosensitivity with chronic hypoxia in four human tumor cell lines. *Int J Radiat Oncol Biol Phys,* 54, 910–920.

zur Hausen, H., Gissmann, L., Steiner, W. et al. (1975). Human papilloma viruses and cancer. *Bibl Haematol,* 43, 569–571.

zur Hausen, H. (2000). Papillomaviruses causing cancer: evasion from host-cell control in early events in carcinogenesis. *J Natl Cancer Inst,* 92(9), 690–698.

Index